Abstract Thoughts: Concrete Solutions

Essays in Honour of Peter Nash

Department of Geography Publication Series No. 29
University of Waterloo

Abstract Thoughts: Concrete Solutions

Essays in Honour of Peter Nash

edited by
Leonard Guelke and Richard Preston

DEPARTMENT OF GEOGRAPHY PUBLICATION SERIES

Series Editor	Chris Bryant
Editorial Assistant and Text Editing	Susan Friesen
Cover Design	Gary Brannon
Cartography	Barry Levely
Computer Systems Design	Marko Dumancic
Text Imputting	Pat Bester Maureen Cairney Douglas Dudycha Jacky Forabosco Susan Friesen Chris Matulewicz Richard Preston MaryLynn Reinhart Susan Shantz
Printing	Graphic Services University of Waterloo

Canadian Cataloguing in Publication Data

Main entry under title:

Abstract thoughts, concrete solutions

(Department of Geography publication series ; no. 29)
"Peter H. Nash publications": p.
Includes bibliographical references.
ISBN 0-921083-26-2

1. Nash, Peter Hugh. 2. Anthropo-geography. 3. Planning I. Nash, Peter Hugh. II. Guelke, Leonard. III. Preston, Richard E. IV. University of Waterloo. Dept. of Geography. V. Series.

GF49.A28 1987 304.2 C87-095082-7

Published by the Department of Geography

Preface

This volume of essays is dedicated to our colleague Peter H. Nash in the year of his official retirement from the University of Waterloo, an institution with which he has been associated since 1970 when he became the founding dean of the new Faculty of Environmental Studies. The position of Dean of Environmental Studies was one for which Peter was singularly well qualified by his academic and professional experience. The new faculty included four academic units: architecture, geography, man-environment studies (now renamed environment and resource studies), and urban and regional planning, and Peter was academically qualified to be appointed to every single one of them. After a distinguished career as a scholar, planner and university administrator Peter had found an academic home that was tailor-made for a person with his multidisciplinary orientation. As Peter himself has written, "I have never worried about boundaries, whether geographical, intellectual, disciplinary, or any other type. One has to follow those avenues where one's intellectual curiosity points the way, even if these paths lead to entirely different territories."

In his academic career Peter heeded his own maxim, but we will not attempt to give a detailed account of it here, because Peter has already done this in his fine autobiographical essay "The Making of a Humanist Geographer: A Circuitous Journey" which is included in this volume. In skeletal summary Peter's academic career included undergraduate and graduate study at the University of California, Los Angeles, separated by wartime service in the U.S. Army. After California Peter spent a year in the Department of Geography, University of Wisconsin before moving on to the Graduate School of Design at Harvard University with John M. Gaus who was one of his Wisconsin mentors. On graduating from Harvard with a doctoral thesis in urban and regional planning, Peter spent several years as a professional planner at the state, federal and local levels before he returned to academic life. After two years with the University of North Carolina, Chapel Hill, Peter became Head of the Department of Geography, University of Cincinnati and then Dean of the Graduate School and Director of the Graduate Curriculum in Community Planning and Area Development at the University of Rhode Island, before finally moving to Waterloo. In 1960 Peter participated in the founding meeting of the International Geographical Union (IGU) Commission on Applied Geography and thereafter attended its many international meetings until 1984, when the Commission came to an end, a victim of its own success. In this period, Peter also attended the famous Delos symposia, organised by Constantinos Doxiadis.

In offering this volume to Peter we do so as a token of the esteem in which he is held by the many colleagues who have had their lives enriched in various ways through their encounters with him. All of the contributors to this volume would have had their own ways of acknowledging their debts, large and small, to a remarkable human being. We ourselves are tempted to talk of the Nash phenomenon. One could indeed elaborate at length on the amazing variety of Peter's interests: in geography, in urban and regional planning, in future studies, in architecture, his many accomplishments as a fair and farsighted administrator, his published contributions to knowledge as a prolific multilingual scholar, the enormous erudition of a person who has rubbed shoulders with some of the intellectual giants of the twentieth century, but one would not in such an exposition really get close to the essence of his personality.

The rich experiences and intellectual achievements might easily in a lesser person have manifested themselves in a certain sense of superiority and aloofness, but not with Peter. His truly engaging quality is his warmth and open-mindedness. In Peter one has a person of vast learning who is always open to new ideas, eager to learn and share his experiences with others on a basis of complete equality. This approach quickly establishes a feeling of trust which permits conversations to range widely. Indeed, with Peter a conversation often becomes an intellectual exploration of difficult or controversial ideas, because his criticisms are open and constructive without ever a hint of "scoring a point" or a suggestion that an opposing point of view is of less value than his own.

The qualities that made Peter an open-minded approachable colleague were also manifested in his classes. As a teacher Peter passed much of his enthusiasm for knowledge and learning on to his students using a variety of teaching strategies with an emphasis on teacher-student interaction. Peter was interested in his students as individuals, and demonstrated this concern in some remarkable ways, one of which deserves a special mention. In the large undergraduate class on the nature of geography Peter gave every student the option of an one hour oral exam instead of the formal written one. Almost every student in this class opted for the oral exam, which says a lot about student confidence in Peter as a person with whom they could communicate without being intimidated, and,

perhaps, even more about Peter who thought nothing of devoting so much of his time to this task.

Peter Nash has touched the lives of all contributors to this volume, and it is entirely fitting that they are drawn from many countries and disciplines. The individual essays cover a broad range of knowledge. There are indeed no rigid boundaries to be discerned here, but we have sought to provide some order to the volume by arranging the essays according to their subjects. In this manner it is possible to discern a certain coherence and logic to the contributions, which are, we believe, in their diversity, a fitting testimonial to Peter's own multifaceted intellectual interests.

* * *

Part one of this book deals with the broad subject of "Planning" and includes four essays. John Mullin considers the gap between theory and reality in city and regional planning. He argues that by focusing on the cultural context in which planning occurs this gap can be narrowed. His approach is demonstrated by a model blending planning theory and reality and by the model's application in case studies. Yehezkel Dror argues that without a new perspective on the theory of planning, significant advancement in its professional practice is impossible. He suggests that the idea of "policy-reasoning" can serve as part of a new perspective. Dror recommends a form of policy-reasoning based on a consistent perspective on action, judgement and practical reasoning. He explores six key issues related to policy-reasoning and demonstrates their application to planning. He concludes that if planners attempt to overcome conflicts between planning theory and practice through the use of policy-reasoning, they must reconsider many of their basic assumptions and techniques, and there must be radical changes in both the professional preparation of planners and in the profession's self-image.

Kenneth Corey describes how Singapore's leaders have planned for an information-age metropolis. Using a "Program Planning Model" framework, Corey traces Singapore's evolution as an information age metropolis through six reasonably well defined stages of development. He concludes that Singapore's experience in both planning and implementing an information based urban system provides insight into issues and problems faced by other cities attempting to develop an information based economy, society, and spatial structure. Francis Horn considers the failure of government planning in South Africa in the face of that country's racial problem. Horn's perspective is that of one individual hoping to learn what might be done to lessen the dangerous state of affairs in South Africa and to avoid further violence. Horn considers apartheid as national policy, the nature of and control of the national government, the dilemma of the Botha government and its relation to punitive measures, and possible solutions to these problems. He concludes that the white minority will not accept a wholly black government in the near future, that blacks must have a meaningful role in running the national government, and that such measures as sharing power and wealth among all races, improvement of nonwhite education, and increasing integration in all walks of life (especially in the schools) must be embarked upon quickly if increased violence is to be minimised.

Part two focuses on "Applied Geography and Planning" and includes three contributions. T.W. Freeman reflects on connections between geography and planning in understanding man-land relationships. He suggests that geographers should have something useful to say about questions of social and economic significance, but cautions that "the value of their comment must rest on a basis of geographical understanding." He demonstrates this position with numerous examples including attempts to predict the future, reconsideration of forgotten explanations of planning problems, the interpretation of landscape, and consideration of major planning problems in Britain and the United States.

Mohamad Shafi defines "Applied Geography," and considers a number of problems facing India, to the solution of which he feels applied geography can make a contribution. Problems related to overpopulation, food shortage, resource management, and the impact of mechanisation on labour are all outlined and tasks for applied geography identified. Particular attention is given to contributions that geographers might make to understanding relations between the distribution of hunger and the structure of agriculture, to population pressure on resources, rural industrialisation, medical geography, recreational geography, transportation geography, and futures research. Shafi concludes that geographers have much to contribute in the search for solutions to many of India's problems. In the last paper in this section, Wolf Tietze argues that demographic studies have placed too much attention on qualitative dimensions. He emphasises the need to focus on the "Quality of Man" in migration research, and uses examples of the extraordinary contributions to American society and culture made by the relatively small number of pre-1945 Jewish immigrants and by the very small number of post-second world war German rocketry scientists.

Part three is organised around the theme of "Geography and Human Values" and is comprised of five studies. Aubrey Diem offers a personal appreciation of the holistic approach used by Peter in his M.A. thesis on *The Vercors Massif: A Geographical Analysis of a French Alpine Region*. Diem emphasises the importance of the "Regional" perspective in geographical education and research and gives special attention to the roles therein of perception, of psychology, and of an integrated view of the world and its landscapes. Highlighting the "Science of Ekistics" approach to the study of settlements and building on recent field work in Cuba and China, Demetrius Iatridis considers the role and importance of human settlements in national development planning. Emphasising a policy approach based on the principles of Ekistics, he compares western market economies with those of Cuba and China. He focuses on relationships between ideology and human settlement function and form and offers an analytical framework (model) for evaluating ideology in the light of Ekistic practices. Considered are level and type of individual participation in government, rural vs. urban settlement growth, and principles of settlement planning. Iatridis suggests that the world-wide human settlement crisis is a derived problem, that its causes are ideological, and that the likelihood of its eventual solution is greater in nations with nonmarket economies.

Gordon Nelson considers the concept of "Utopia," defines it, notes its place in interpretations of human environments and considers examples of four types of utopia: mountains and wilderness, the oasis in the drylands, the world city, and the utopia for every man, the vernacular or ordinary landscape. Nelson emphasises several meanings of the term "landscape" and by his four examples: Banff National Park, The Cypress Hills, Paris, France, and Point Pelee, Rondeau, and Long Point Peninsulas, demonstrates advantages of thinking about our surroundings, past, present and future, in landscape terms. He suggests that most ideas of utopia are in fact very old, that the examination of utopian ideas and of landscape in utopian terms tends to highlight and initiate dialogue on much that is thought to be good for human existence. He notes that utopian concepts of nonwestern origin should be studied more thoroughly, that consideration of utopian landscapes tends to bring thinking about humans and nature closer together, and, finally, that consideration of utopian landscapes draws thinkers into the realm "where dreams can meet with reality."

Leonard Guelke in a comparative study of the frontier examines three interrelated themes of frontier settlement in North America and South Africa. His themes include the factors generating frontier settlement, the nature of frontier society and the long-term ramifications of the frontier experience. The "motor" of frontier expansion, he argues, was not free land as such, but rather the prospects frontier areas held out to settlers of an independent livelihood. This livelihood was typically based on near subsistence agriculture and hunting and early frontier peoples had very limited commercial contacts with the outside world. The large degree of economic independence encouraged the widely-celebrated values of independence and individualism among frontier settlers. The conditions which encouraged near subsistence agriculture and social independence, Guelke suggests, did not long endure in most regions of North America, but they did persist in South Africa and were of great importance in shaping the values of generations of white settlers in that country.

This section concludes with Ludwick Straszewicz's perspective on "Forced Housing." He argues that alongside the free world experience, in which most people are free to live where they wish, there is another world where political power decides where people live, often against their will. Straszewicz considers examples of forced housing from the middle ages, the Jewish Ghetto, and as a result of discrimination against immigrants and others perceived as "different" on social and other grounds. Relations between forced housing and forced migration and natural boundary changes associated with World War II are considered along with the present racial situation in South Africa. He suggests that in the world today most people live under conditions of forced housing, and that although forced housing is present in democratic countries, it is closely connected with autocratic political systems and with authoritarian regimes. He suggests relations between forced housing and social and economic environments as a promising research area.

Part four is devoted to the subject of "Urban System" and includes three papers. Richard Preston and Clare Mitchell suggest that economic base and central place theories can be combined into a single overall explanation of the functional dimension of urban systems. They show that the theories are not equivalent in the range of activities embraced by their respective domains or in their ability to generate normative activity or settlement patterns. Nevertheless, they found important areas of equivalence between the theories that provide a base for their combination, and they suggest that equivalence between the theories can be increased by expanding the domain of central place theory. Christian Dufournaud contributes to a solution of the problem of updating input-output tables. He presents and reviews critically a number of techniques used to solve this problem and presents some updates of the 1967 Washington State table to 1972. He concludes that this preliminary analysis was fruitful enough to make follow up analyses worthwhile. James Bater considers Soviet urban development in terms of "first principles and present realities." He reviews some of the more important Soviet urban policies and their impact on the size and distribution of urban places and comments briefly on possible urban development trends during the Gorbachev era. After noting that establishment of the "Soviet Socialist City" offered a new opportunity for charting the course of Soviet urbanisation and reviewing the debate over future urban form that took place early in the Soviet period, he considers conditions that became general principles for urban development in the USSR: mechanisms for manipulating population movement, investment, and location of new facilities - especially in industry. Within this context, special attention is given to labour management, attempts to promote growth in specific urban size categories, and to attempts to plan the national settlement pattern according to urban system concepts. Bater suggests that Soviet policies designed to check the growth of the larger urban places and to promote the expansion of the smallest places have not been especially effective and that in the USSR the relationship between planning principles and practice in the pattern of urban development is not necessarily a close one.

The volume concludes with two appendices. The first presents Peter's autobiographical essay "The Making of a Humanist Geography: A Circuitous Journey" published earlier in *Geography and Humanistic Knowledge, Waterloo Lectures in Geography, Volume 2*, edited by Leonard Guelke (1986), while the second appendix presents a list of Peter's published works.

Leonard Guelke

Richard Preston

Department of Geography
University of Waterloo
Waterloo, Ontario

Acknowledgements

The editors would like to thank the Department of Geography and the Office of the Dean of Environmental Studies, University of Waterloo, for financial assistance toward publication of this *Festschrift*. Acknowledged also are Pat Bester, Jacky Forabosco, Chris Matulewicz, Richard Preston and Susan Shantz, with special thanks to Maureen Cairney, Douglas Dudycha and MaryLynn Reinhart for text inputting and to Susan Friesen who prepared the manuscript for publication. The computer program was designed by Marko Dumancic. The cartography was prepared in the Environmental Studies Cartographic Centre under the direction of Gary Brannon.

TABLE OF CONTENTS

LIST OF FIGURES

CHAPTER 1

CHAPTER 9

CHAPTER 13

CHAPTER 15

LIST OF TABLES

Abstract Thoughts: Concrete Solutions

PART I

PLANNING

CHAPTER 1

THE PLANNER AND LOCAL CULTURE: A JOURNEY INTO FOREIGN TERRITORY

John R. Mullin
Department of Landscape Architecture
and Regional Planning
University of Massachusetts at Amherst
Amherst, Massachusetts

INTRODUCTION

Planning theory as a subject of scholarly inquiry is fraught with pitfalls. It is necessarily abstract but without the bounds of other academic disciplines. It must reflect reality and yet the reality is so broad that efforts to cluster experiences into meaningful concepts can never be complete. Still the quest continues. Professors who teach planning theory tend to lecture on those topics in which there is some boundedness. These include the planning process, value clarification, means and ends, choice theory, role alternatives and ideology. When they are taught, they tend to be presented in as standardised or sanitised manner as possible. By so doing, instructors can be coldly analytical in, for example, comparing Toronto with El Paso or even the Annapolis Valley with the Pioneer Valley. The net result is that instructors and students gain a sense of the broad, fundamental principles that guide planning theory, but rarely any understanding of how theory relates to actual practice. In short, theorists are doing reasonably well teaching planning as an abstraction. They (we) are not doing as well concerning the relation of abstraction to what actually happens in the planner's world.

What can be done to correct this shortcoming? While there are many potential options, to this author, the most important missing ingredient in the teaching of planning theory is our inability to instill knowledge in our students concerning the cultural context in which planning occurs. I base this position on my ten-year struggle as an instructor of planning theory and as an active planning consultant in more than 150 communities across New England. As a lecturer, it became clear that my subject matter was incomplete. As a practitioner, looking for contextual help from theory, I also found that something was missing. With these impressions in mind, I have begun to develop a component in my theory courses that hopefully will help overcome this shortcoming. It focuses upon the cultural context in which planning occurs and teaches basic cultural principles that planners must grasp if they are to be effective. How this occurs is presented below.

This study focuses upon Massachusetts, a small state in New England, that is at once Catholic and Protestant, conservative and liberal, international and provincial, traditional and modern. Attempting to grasp the context of the state is like grabbing jello - it simply cannot be done. The cauldron that is Massachusetts makes it an exciting place for a planner to work. At the same time, because of its unique elements, the experience does not travel well. There are still lessons to be learned.

There are three major parts to this paper. It begins with an explanation of a simple model that defines the four cornerstones of planning practice based on my own experience. From there, a series of case studies are offered as examples of how culture influences planning. Finally, this paper closes with a set of summary thoughts on what the experience means to planning as a body of knowledge.

A SIMPLE MODEL

Throughout my years as a consulting planner I have been constantly struck by the different perspectives of the citizens in the towns in which I have worked. After arriving in town, comments would be made such as, "*Now* this town can work like a business," "Great, now we can get our share," or "Finally we can keep our town as it should be." In each case, there was a perception on the part of the people that somehow the planner would come to town, diagnose problems and create solutions that could be rapidly implemented. After hearing such comments and noting the perceptions, I always came away disappointed because I know that these are expectations that would rarely match reality. The planner, more than anything else, has to balance competing interests, ideologies, and laws to the point that optimal solutions are rarely approved. It is this sense of balance that has led me to construct Figure 1.

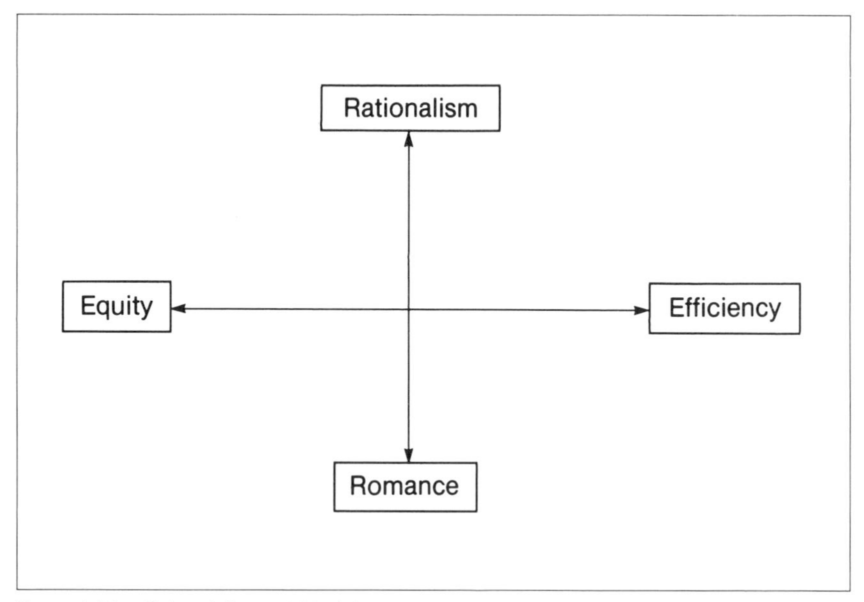

Figure 1: The Cultural Context Model

There are four elements. The first, *Rationalism*, means simply that the community will make deliberate and thoughtful decisions concerning its future. When the question is survival, there is little doubt that rational decisions will be made. For example, if water must be protected for the well being of the community, there is no doubt that a rational decision will be reached: either it is protected or it isn't. Here there is a clear understanding of the problem (the threat to the water supply) and an understanding of the range of acceptable options. The planner's job, simply stated, is to match the problem with the desired option and oversee its resolution. In many ways this is the easiest type of problem for the planner to correct.

The second element, on the opposite side of Figure 1, is *Romance*. Here concepts such as self-interest, love of the land, territoriality, class awareness and preservation are important. These values, when strongly held, can cause havoc to the planner faced with critical, yet less than survival, issues. The romantic typically has a fortress mentality and accepts change grudgingly. For example, when higher government imposes change on the local community, the people will inevitably organise and resist. A recent example of this phenomenon in Massachusetts involves the expansion of a state highway. This highway, if expanded, would have helped to improve the economic vitality of a depressed region. Yet it would have also destroyed part of a forest that is considered a local treasure. The local town wanted no change, the region and state wanted major change. The town, through its political representatives, conservation and environmental action committees and its own organising efforts effectively blocked the project. Who won? Whose public interest was paramount? Clearly, the planner in this context must operate quite differently than the planner working under rationalism.

Neither rationalism (as the technical decision mode) nor romanticism (as the ideological decision mode) tell the complete story. There are also the elements of Equity and Efficiency. *Equity* in this context refers to the improved distribution of society's goods and services to the point that the basic needs of all citizens are met. While one can argue about the scope of basic needs, the necessity for food, shelter, security, education and opportunities for jobs would fit all definitions. In Massachusetts, the equity issue is being faced most dramatically in terms of housing. Should every community provide a range of housing for all economic classes of citizens? In theory, few would object. In reality, it is one of the most difficult problems facing the planner. The state courts and the state government are forcing local communities to provide housing for all income groups. The premise here is that no town is an economic island and all communities must provide their fair share of housing for all income classes.

Efficiency means providing the maximum amount of services for the lowest price and minimising the presence of government in the daily lives of the citizenry. In a planning context, it means planning under capitalism - does the decision result in an increased tax burden and/or does it expand the tax base? Or does the decision mean that the citizen will be able to live with little government intrusion? We see this most vividly in terms of a reluctance to invest in the future. Massachusetts now has one of the lowest rates of long-term indebtedness in North America. As a result, infrastructure systems (roads, water systems, sewer systems and capital facilities) are eroding rapidly. Further, the passage of a tax cap on property revenues means that there are now severe limits on how much revenue can be raised locally. There is now a belief in the state that the government that governs least, governs best. The political liberalism that once marked the Massachusetts electorate is, for the moment, in full retreat.

In sum, regardless of where on Figure 1 a decision is placed, there is a potentially large group of citizens, given pluralism, who will be disappointed. The saving grace is the fact that all decisions are made in a cultural context, such as that represented by Figure 1. As long as the planner understands this context, relates to it in his/her work and employs it in decision making, some progress can indeed be made. This situation is emphasised in the cases outlined below.

THE CASE STUDIES

The Catholic Church Meets Harvard University

Over the past decade, there has been extensive interest on the part of Harvard University, among other greater Boston universities, in genetic research. At times this research was "pure" in the sense that it was research for the sake of research. However, at other times, the research was "applied," meaning that experiments were being undertaken on organisms with the intent of influencing the genetic code. It is this latter action that involved the Cambridge city planners.

The Cambridge City Council, largely blue-collar and Catholic, was profoundly disturbed about this genetic interference and began discussions concerning the efficacy of this research within its territory. The city could indeed control this research under its health and safety police powers (what would happen if "it" escaped?). It also had the right to control it under the "morals" clause of its police powers.

The timing of the discussion concerning this research coincided with two other issues of public protest: the responsibility of medicine and science in nuclear research and the question of legalised abortions. All three issues reached the streets and were increasingly emotionalised. Thus we see a predominantly Catholic City Council, with a strong anti-Harvard bias, trying to decide whether or not academe could undertake research within its own confines unfettered by local control - a role universities across the globe have attempted to maintain for centuries. We also see a group of local citizens concerned with local issues and motivated by ideological principles that run counter to the research in question.

In a rational sense, the issue could be easily resolved: Harvard could simply police its own research. After all, it has an international reputation of the highest order. And yet, Harvard, within Cambridge, is perceived as a foreign entity often preaching values that are little understood or accepted.

Thus we see an issue that, say, in a typical midwestern setting could be easily resolved, becoming constantly escalated to the point that is a moral issue. It also had to be handled with the wisdom of Solomon. And so, for the moment, it has. The research is being allowed albeit with increased city monitoring. It will continue to simmer until an accident occurs or the citizenry become aroused again.

Where does this problem fit on Figure 1? Clearly Harvard would have liked it to fit at the top of the Rationalism line (Harvard would police its own research). The City of Cambridge would have liked it to rest near the bottom of the Romance line. The fact that the decision, for the moment, represents a compromise and is therefore near the centre point means that neither side was satisfied (Figure 2). And so the decision stands.

Concerning the role of culture, this entire issue was strongly coloured by local conditions and local values. The fact that the City of Cambridge is an Irish-Catholic city of blue-collar workers, has a strong tradition of nitpicking at Harvard, has several enclaves of strong citizen participation and active political organisation, and that it was involved in an issue that had ramifications across the nation, created a cauldron in which unique ingredients created a one-of-a-kind stew.

Could the planner have entered into a rational discourse on this subject? Could the planner have served as an agent of change in which he/she attempts to modify the beliefs, attitudes, and values of the people? The answer is no, this could not be done. The environmental and cultural circumstances that make Cambridge what it is represents the context in which the planner must work.

The Politicisation of Bureaucracy: Mayor White's Planner Meets Mrs. Jones

In the mid-1970s, Kevin White, then Mayor of Boston, began to realise that the city bureaucracy was not functioning as it should. Under normal circumstances, if Mrs. Jones had a pothole on her street, she would call the Department of Public Works (DPW). The DPW would then make out a work order, and in time, dispatch a crew to make the repair. The Mayor, after examining his bureaucracy, believed that this system was in utter chaos, and because of civil service and years of patronage, not likely to change. After reviewing how the Mayor Daly Machine operated in Chicago, Mayor White brought the Chicago model to Boston. Under this model, every city

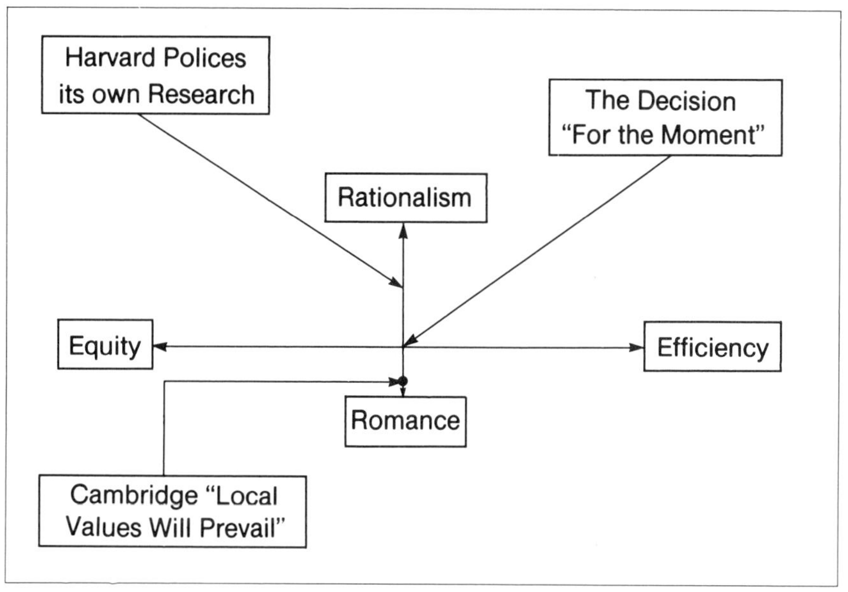

Figure 2: The Catholic Church Meets Harvard

employee, including the professional planning staff, was given an additional "political" responsibility. For example, Planner Smith, responsible for planning in West Roxbury, was told to report to the West Roxbury Ward Captain who, in turn, gave him a list of names that were his responsiblity. Henceforth, if any of these families had a problem with the city bureaucracy, they were to call Planner Smith directly. Planner Smith, working through the Ward Captain, would then see that the problem was resolved. Once it was corrected, Planner Smith would stay in touch with Mrs. Jones and her neighbours, always reinforcing the point that Mayor White, not the City of Boston, was responding to their needs. At election time, Planner Smith would work overtime to insure their support for the Mayor, even taking some vacation hours on election day and escorting his people, if needed, to the polls. His pay raise was determined by not only how good a planner he was, but how by his people voted. Incidentally, there was considerable exit polling to determine how well he did.

What are the issues here? There are two. The first is the ethical issue concerning Mayor White's politicising the bureaucracy. Clearly, in terms of Civics 101, the system was supposed to work. Civil servants are hired to meet the needs of the "civitas": free men and women performing labour for the public good. The fact is however, that not all people are motivated. When they are not, the system fails. In private enterprise these people would be fired. Under Civil Service, they can only be fired for a proven cause - an almost impossible task to prove and one that requires massive hours of public hearing time on a case-by-case basis. Simply stated, a civil servant is virtually never fired for marginal or slow performance of duty. What is the mayor to do? Clearly changing the rules governing civil service and increasing working accountability would be an appropriate step - albeit a step that would take years. And what would happen in the interim? Clearly publicising poor performance would also help. On the other hand, the workers would know that action would be all the mayor could do. In the long-term this would show that he was little more than a paper tiger. In the final analysis, the choice was to do nothing for the remainder of his term and let the delivery of public services deteriorate, or he could politicise the bureaucracy and insure increased performance. He chose the latter. Was Mayor White's decision unethical? If he was Mayor of Dubuque or Omaha, his actions, most likely, would have been considered highly suspect in that cultural context. On the other hand, in Boston, where politics is a daily sport, it is hard to condemn his actions. Yes, one could condemn him for using city employees for his own political gain. But then again, he took the only step he could to improve the responsiveness of the city bureaucracy. Certainly in the context of the local politics this decision was in keeping with local cultural traditions.

The second issue relates to the planner, a professional who values his/her white hat, who values his/her "above the bureaucracy" position, and who is governed by a professional code. Is there loyalty to the city, the people, the mayor, or the professional code? In a totally romantic sense, since the planner pledges fealty to the public as the civitas, he/she would quit. However, he/she too knows that the bureaucratic system is wrong and unchangeable. Further, he/she, typically with mortgage and family, knows the turmoil that comes with unemployment. Thus we see the planner accepting the politicisation of planning and responding accordingly. And, by so doing, further strengthens the cultural context that accepts this type of behaviour.

Where does all of this fit on Figure 3? The planner's role would fit somewhere in the upper left quadrant, between rationalism and efficiency. The rationalism would be represented by the pragmatism of the planner's decision, given that quitting was out of the question. The efficiency would be represented by the ability of the planner to provide a better, quicker product without bureaucratic impediment. Equity would not be represented due to the fact that only those in service to the mayor would gain. And romanticism would not apply for clearly the democratic ideal was not being met. Above all, it was the local political culture that dictated how decisions were made - for better or worse.

The Planner Meets The Contras

In the summer of 1984 the United States Army National Guard contracted with a consulting firm to develop a master plan for the long-term development of Camp Edwards Military Base on Cape Cod. The plan, complete with proposals for forty million dollars worth of new offices, barracks, administration facilities and new ranges, was completed in the fall. It only required approval of the Governor and the state's congressional delegation before funds from the federal government would be released for the plan's implementation.

Surrounding the base were the towns of Sandwich, Bourne and Falmouth. These towns, with minimum zoning controls, boomed through the 1970s and 1980s. Part of the growth occurred immediately adjacent to the base. Anyone who has lived near a military base knows the impact of sound and vibration when howitzers are fired. Nevertheless, new home owners continued to move into the area.

In the fall of 1984, the plan was released to the public. Immediately, an uproar occurred with public hearings, editorials, and radio talk shows fueling an extensive debate concerning the base's future. Should it be expanded or should it be increasingly restricted? The Army stated that it had been using the base since the 1930s and had the right to expand to meet its mission requirements. The new residents protested that the firing of weapons was destroying their quality of life, causing excessive vibrations and contributing to noise and smoke pollution. When advised of the principal of "Caveat

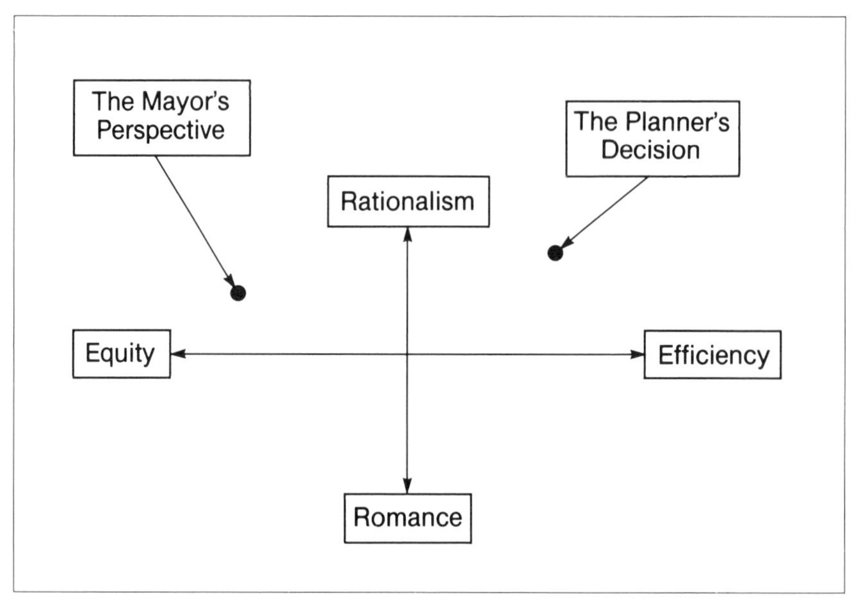

Figure 3: Mayor White's Planner Meets Mrs. Jones

emptor," these people were not amused. It should also be noted that the Army does not vote in local or state elections.

These citizens joined with environmental organisations and began formally to protest the expansion of the base. Like most sophisticated advocate groups, they delved into their legal rights and the detailed contents of the plan. Upon reading the plan, the citizens noted that the plan included new facilities for the training of "Special Forces" units. (These troups are currently most identified with the American counter-insurgency in Central America). The recognition of this fact brought an explosion of protest, including blocking the entrances to the base, picketing on the steps of the State House, and presenting petitions to elected officials. Their actions regularly made headlines in the Boston and Cape Cod newspapers. The Governor, a political liberal and an advocate of staying out of Central America, quietly supported the citizens' position.

In the spring of 1985, the Governor, as Commander-in-Chief of the National Guard, announced that the plan would not be approved, stating that a full environmental impact statement (EIS) would be first required before he could decide. An EIS of this type normally takes three years to complete. This decision also meant that the designated federal funds would be allocated to some other military post. At the same time he announced that Massachusetts troops under his control would not be allowed to train in Central America and that no facilities would be built at Camp Edwards that contributed to Special Forces training. The protestors clearly won and while doing so they even contributed to a crisis at the national government level concerning who controls the state militia. The plan at the moment is "dead in the water."

This is a case where ideology, self-interest and political power combined to influence a decision. The basis for the decision did not really concern the master plan. Rather, the Governor's position was based on his opposition to the President's Central American policy.

Where does this decision fit in terms of Figure 4? It was not rational in the sense that the decision was based upon the merits of the plan. Equally as clear, the decision did not involve issues related to equity or efficiency. The Governor's stance seems to fit in the romantic category - the denial was a statement that *his* troops on *his* base would not contribute to the implementation of policies with which *he* disagreed.

What is the lesson here? At one level, it is that the range of influences that govern the acceptability of a plan do not always rest within the context of the plan's merits. There are also external inputs to consider. All of the planners involved in this plan were stunned when they learned that the plan's acceptability was to be determined by its relationship to national policy. Secondly, planning is a political activity. As such, the planners must have knowledge of the political context in which decisions are made. A governor who is against United States policies in Central America and who is interested in running for higher office is not going to support a continuation of this policy in an area that he controls.

The Planner and Paternalism

Throughout the last twenty-five years, the City of North Adams, Massachusetts, was marked by its tired old textile and paper plants with one exception - the Sprague Electric Plant. Well managed, sophisticated and profitable, this company expanded throughout this period until by the 1970s, it was the city's major employer and most powerful political force.

As the company grew, local planners and politicians endeavoured to stimulate spin-offs and attract similar plants to the area. Unfortunately, each time an attractive potential employer came to the

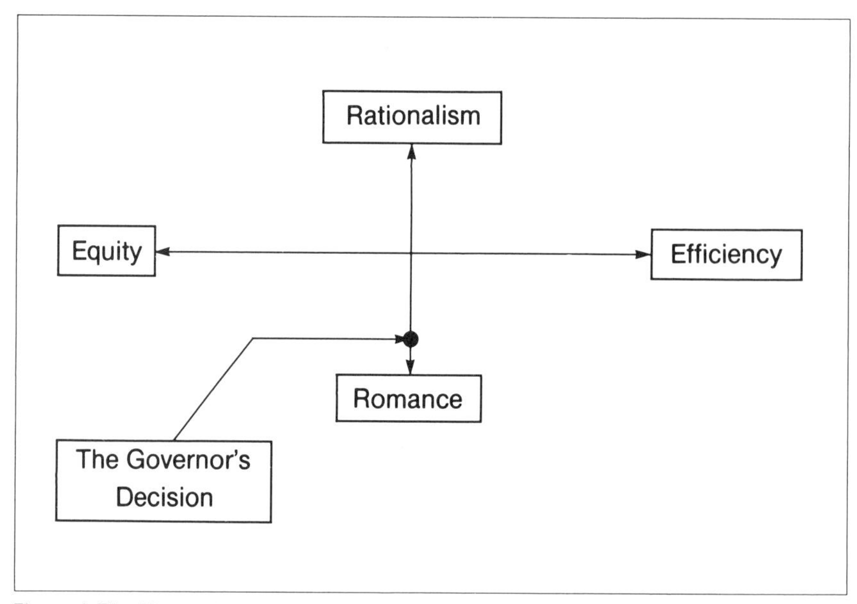

Figure 4: The Planner Meets the Contras

city, the word spread from the company that if a new firm relocated in North Adams, the Sprague Company would move out. Fully realising that the company was already expanding elsewhere, the city knuckled under and decided to live or die with the company.

In the early 1980s, the founder's grandson took charge of the company and immediately made plans to relocate the overwhelming share of its operations. Over the next two years, more than 1,200 workers were laid-off and only a handful of the rank and file were offered positions in other plants. North Adams today has the highest unemployment rate in the state, is losing population and its long-term prospects for recovery are weak.

The planners in this case relied strongly on the private sector and endeavoured to employ corporatism as a means of insuring that both local needs and the needs of the Sprague Company were met. This implied commitment on both sides. The city kept its side of the bargain and insured that competing firms would not locate in its confines, and by so doing, insured a captive, low-paid and "grateful-for-jobs" labour force. Unfortunately, the bargain was toothless: over time, the city needed Sprague more than Sprague needed the city.

In terms of Figure 5, one could place the city's decision to support the Sprague Company somewhere between Romanticism and Equity. The city was responding to the needs of its workers, the company was a good citizen and it was a long-term resident. Loyalty on the part of New England workers dies hard. From the company's perspective, the decision to move should be placed on the direct opposite side. It was made due to the need to be closer to markets, increase profitability and increase efficiency (mainly by shedding the local union).

If anything, this case shows the difficulty of responding to paternalism. The economic base of the city and the well being of the company went hand-in-hand. And yet the city was powerless in terms of what it could do: decisions were made by an entrenched hidden hand consisting of company-connected people. Clearly, if the planner was to be successful by any measure, his/her actions would have to meet the approval of this group. And so it was.

There was an underlying theme in this case related to acceptance of the status quo. Our organisational theorists teach us that through education we can change the beliefs, attitudes and values of the public. In effect, the message is that planners know more than the citizens about the consequences of actions. In some cases we do, in other cases we don't. It seems, however, that to move from the abstract thoughts to concrete results, there is a need for commitment, consistency, coherency, and a long-term vision on the part of the planner. At the local level these characteristics do not exist. To begin with, the average life of a planner in a New England community is less than five years. As well, the linking of planning to politics means that the results of planning are tied to the political term of elected representatives. One increasingly hears the phrase from planners, that "if it can't be done in one term, it can't be done." Finally, when the choice is a fair job now against a good job later, the public will inevitably support the former. The future in our communities has now been compressed to less than a five-year time horizon. Within this time frame, the opportunities for complex educational strategies of any type are almost nonexistent.

"Johnny Appleseed Goes Condo"

In 1985, the planner for the City of Leominster, Massachusetts, called and asked for help concerning a housing development proposed to be located on one of the last remaining apple orchards in the city. He explained that Leominster was the home of Johnny Appleseed, an

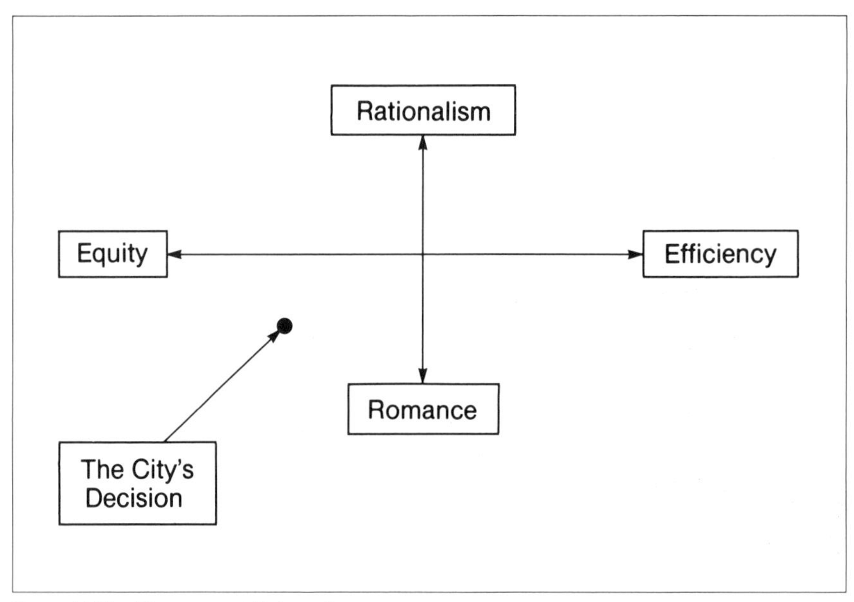

Figure 5: The Planner and Paternalism

American folk hero who allegedly spread apple seeds across the northeast. He also explained that he wanted an alternative created that would allow the development to be built while saving the orchard.

After examining the site, it was determined that a condominium project could be built in such a way that the orchard could be saved. The developer was supportive of the idea. However, the grouping of the structures would require a higher density than allowed under the local zoning and, therefore, City Council approval would be required. Two proposals were put forth to the City Council. The first showed a typical gridiron layout of houses that destroyed the orchard. The second showed a condominium development with the orchard intact. At the public hearing, presentations were made that stressed the need for the housing units, the saving of infrastructural costs and the fact that the orchard would be saved. The neighbours attacked the positions, stating that "outsiders" would be living in these "glorified apartments" and that their quality of life was being threatened. The neighbours won on a 9-0 vote. Today, the houses are built, the orchard is gone and the neighbourhood, by all opinions, has deteriorated dramatically.

This decision would fit someplace between Equity and Romance (Figure 6). Equity in this case would be determined in terms of "fairness" to the neighbourhood. The Romance element consisted of the self-interest of the neighbourhood in terms of protecting their properties. Clearly, neither Efficiency nor Rationalism, from the City's perspective, was important.

A critical point in this case and in the others presented in this paper is that planning decisions are made based upon the impact of the person, house, neighbourhood or town closest to the project. At no time in this project did the City Council ask the question "What does this project mean to the City as a whole?" Instead, they made their decision based upon the feelings of the adjacent property owners.

Leominster is a blue-collar city. Its workers to a large extent consist of first-generation Americans who are pursuing the dream of owning property. As such, property is to be protected at virtually all cost and anything that threatens its value must be fought against. It is this cultural context within which the planner was attempting to bring change. In retrospect, there is little wonder that the ideal failed.

POINTS TO PONDER

In each of the cases presented above, what do we see in terms of the planner and the culture in which he/she operates? Above all, we see something about which he/she has had no training. For example, we teach process from a democratic perspective. This process calls for the diagnosis of a problem, the development of alternatives, the selection of one approach, and ultimately, the implementation of corrective measures. Such a system works well when the planner is dealing in the technical realm - where the choices are finite and backed by scientific rules of evidence. Land use decisions, for example, are at least partially dictated by soils, slopes, permeability and ground cover. A lot on a fifteen per cent slope with a clay lining is not developable under normal circumstances. But increasingly the actions of a planner are outside of the technical realm and now involve a new set of rules. What are these rules?

First, to paraphrase the former Speaker of the United States House of Representatives, Tip O'Neill, *all planning is local*. The federal government and the state government can propose, but it is the locality that will dispose. Nothing illustrates the point more vividly than the state's current decision to force communities to accept their share of the state's poor by providing affordable housing. This

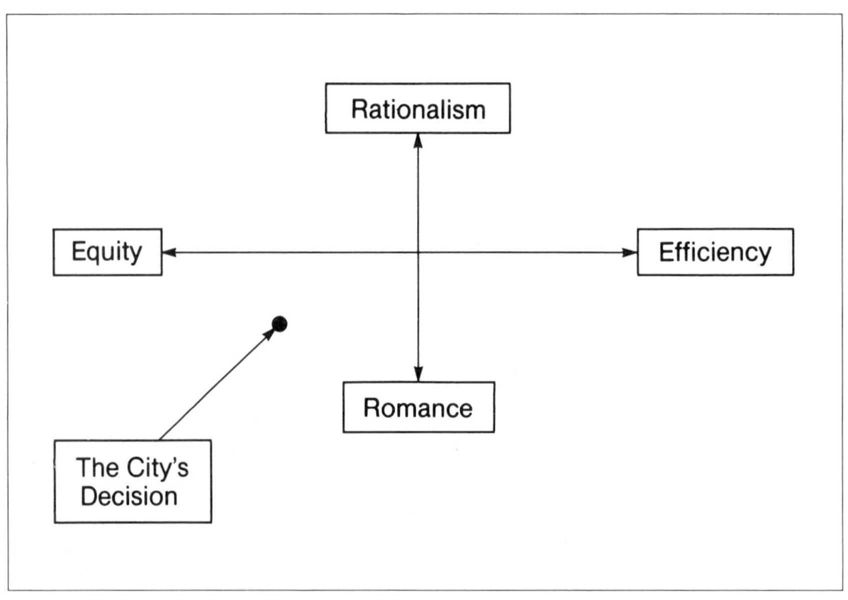

Figure 6: "Johnny Appleseed Goes Condo"

ruling is anathema to local citizens who, bluntly, do not want to accept outsiders who haven't earned the right to move in. Yes, racism and class awareness are involved. However, not to the degree that one would expect. In most cases, minorities would be welcomed provided they paid their way. The net effect is that towns are paying lip service to the law. They are meeting its letter through providing a token number of units and placing them in areas where there is no local resistance. Yes, there is compliance, yes, there is local action, and yes, there is an increase in the aggregate supply. But in no case has the acceptance meant an integration of these units or people into the community. These people live in "the project" or are "government subsidised." They may live in the community but are not of the community. Hopefully this will change. Unfortunately, the evidence to date is that it has not.

Secondly, *all planning is political*. This applies, to a degree, even to the interpretation of technical planning activities. For example, I was recently involved in a rural town that had no sewer system but was increasingly subject to growth pressures. The townspeople were split over whether this was beneficial or not. The "no-growth" Board of Selectmen realised, given the split, that the town meeting would not support any further restrictions. They realised, however, that they had the power to determine the rates at which percolation on the land occurred (the faster the required rate of absorption, the less the potential for development). By administrative fiat, they doubled the rate at which the land had to percolate and therefore were able to slow the pace of development. The point here is that there is little separation of technique from politics.

Third, *there is no such activity as "value-free" planning*. Whether it is a cause against the Contras or in support of high tariffs, the fact remains that decisions are rarely reached based upon scientific evidence. As a case in point, we were recently involved representing a developer to a small New England town. The developer wanted to place clustered condominiums on a parcel that perfectly fit the density and use requirements in the town's zoning ordinance. However, clustering requires that a special permit be granted by the town council. The proposed project showed nothing but gain for the town (taxes, needed "starter" units, beautiful design and a donation of land for recreation). It was overwhelmingly voted down. A majority of the town council saw condominiums as "apartments" and felt that apartments would attract undesirables. Our fiscal impact analysis, housing needs analysis and plans showing how the project related to community character were ignored. Their minds were made up the minute the word "condominium" was put before them.

Fourth, *all planning is short-term*. In other words, one can worry over long-term OPEC policies, but until these policies impact gas at the pump, there will be little response from the public. We see this vividly concerning social service decisions. The federal and state governments have developed policies designed to deinstitutionalise prisoners, addicts, and the mentally ill. When the decision is at the policy level, one finds little negative response to the idea. However, once implementation begins and the actual sites of the group homes are being selected, then the fact that a site is part of a national policy or a state strategy is meaningless and the NIMBYs (Not-In-My-Back-Yarders) become active. It is only when decisions reach the impact level that they can be understood by the average citizen, and that the success or failure of a plan can be determined.

Fifth, *dramatic change is distrusted*. Nowhere can we see this as clearly as with reindustrialisation. In town after town that has lost its industrial base, we regularly note that the local citizens wish for more of the same types of industries that recently closed shop. In most cases, given choice, they would prefer a textile or paper factory. This is despite the fact that Massachusetts has been shedding these plants for decades, that they offer low pay and that there are few opportunities for advancement. There is "comfort" with these firms because they match the skills of the displaced workers. Given a choice of dramatic change or no change, citizens will usually opt for more of the latter than the former.

Sixth, *equity must be interpreted in terms of local needs.* When informed of the need to provide help for the elderly, the young and the homeless, there is little resistance at the local level. These people, from the local citizen's perspective, are "of the community" and therefore are a local responsibility. It is when equity is interpreted on a regional basis or statewide that there is concern. There is fear that if it is determined on a regional or larger basis, they would have to accept "outsiders." A key point here is that *territoriality is strong at a municipal level.* There is a sense of home, of belonging and knowing one's place, that is strongly held even if the resident has been living therein for only a short while. This same feeling does not extend to the region or to the state. The military base case study discussed earlier represents a critical example of this phenomenon in the sense that newcomers living on the fringes of the base were, in effect, defending their turf against outsiders - in this case the United States Army.

The seventh point is that *local elected officials rarely have control over town matters.* The presence of paternalism, the hidden hand and/or a competing power base is such a frequent occurrence that, in the more than 100 cities and towns in which I worked, less than ten per cent of the the time have I found local officials who are in charge of their community. Quite often, one must visit the country club, the board room or the local bar in order to find out who really is in charge.

Finally, *the idea of the planner as one who educates the public is rarely accurate.* We have been taught that it is the responsibility of the planner to inform the public about specific problems, alternatives and best solutions. Somehow, through careful discourse and debate, the public will then make the best choice. It rarely happens that way. First, the planner is typically an outsider with little commitment beyond professionalism to the community. Second, whether the planner likes it or not, he or she is part of the bureaucracy. This makes the planner automatically a nonobjective person. Third, the planner typically speaks a language that is only partially understood by the public. It is axiomatic in Massachusetts that if the local planner takes a stand at a town meeting then the public will go against his/her position. Education only works when the planner is able to make other officials or citizens the educators. In effect, any plan to be effective at the local level must become the citizens' plan. If this does not occur, then planning efforts will be largely in vain.

SUMMARY

This paper is written more as a polemic and a plea than anything else. As an instructor of planning theory and as a practitioner, I cannot help but note the frustration on the part of students who struggle to see relationships between abstractions and reality. At the same time, practitioners regularly wonder how their activities relate to the professional body of knowledge. There is a clear need for increasing emphasis on the relation between local planning and higher levels of strategies, policies, and theories. Until we obtain a clear understanding of what happens on the ground, there will be little narrowing of the gap between abstraction and reality.

Finally, there is need for our planners to gain skills in understanding the cultural context of planning. Planning is not only cost-benefit, land use modelling and impact assessment. It is also the ability to manage change in a manner that enables citizens to maintain and enhance the quality of their lives. If this is to occur, one must first understand who the people are and what their values are. All planning actions follow from this fundamental step.

CHAPTER 2

PLANNING AS A MODE OF POLICY-REASONING

Yehezkel Dror
Professor of Political Science and
Wolfson Professor of Public Administration
The Hebrew University
Jerusalem

THE IDEA OF POLICY-REASONING

The literature on planning is expanding all the time, but few new ideas on planning in the generic sense, as distinguished from monographic information on particular planning experiences and from disciplinary content related to particular domains of planning, can be found in it.[1] The same is true, though to a somewhat lesser extent, in more related methodology-oriented domains of decision studies, such as policy analysis and systems analysis.[2] My own explanation for this

---

[1] Partial exceptions are illustrated by Jennergren, 1978, Faludi, 1973 and 1986, Healey, McDougall and Thomas, 1982. Innovative are some papers presented in Morley and Schachar (eds.), 1986. But these books also illustrate the limits of planning theory within its present paradigms - as do the good surveys of planning literature and approaches in Wilson, 1979 and 1980.
It is interesting to ponder the idea that much of actual planning practice is better than its reflection in literature dealing with planning theory, as is also the case in much of policy analysis and, comparably, in much of public administration. Better explication and processing of valuable planning experiences and tacit planning knowledge may well be a good way to advance some parts of planning theory. This possibility is not examined in the present essay.
[2] On the distinctions, if any, between "planning," "policy analysis," "applied systems analysis," etc. and their dependance on political

state of planning theory is that an appropriate frame and base is missing and that without quite different perspectives on planning, significant advancement of its professional practice is not possible.

The idea of policy-reasoning can serve as one possible foundation for advancing planning. On the most fundamental level, a philosophy of policy-reasoning is needed, which is based on the philosophy of action and of judgment and on practical-reasoning,[3] while meeting the particular features and characteristics of policy making. The present tendency to base planning and related approaches on a diluted idea of philosophy of science needs rectification, because of significant differences between the appropriate criteria of "scientific knowledge," whatever they may be at different periods and in different disciplines, and the criteria fitting an action-oriented professional endeavour such as planning. For instance, philosophy of science recommends admission of ignorance til a validated finding is available which meets abstract criteria of "truth;" but in planning, because of the unavoidable rhythm of decision making, including the default choice of "not deciding," the criteria of "preferability" are more appropriate, the relevant question being whether actual action will be better with the help of feasible planning inputs than without them.

Leaving the deeper philosophic aspects of policy-reasoning to another opportunity (Dror, 1986, pp. 221-24 and Dror, in press), this paper pays tribute to my friend and colleague Peter Nash by exploring

----

language games, see Dror, 1986, pp. 162-65. My own tendency is to group all these policy oriented prescriptive endeavours together and to regard them as one cluster, which can if desired be subsumed under the term "policy sciences."
Whether this approach is accepted or not, as a matter of fact policy analysis and systems analysis literature at present is stronger in methodological innovations than most of planning literature. Still, the prevailing weaknesses of policy analysis and systems analysis are serious, reinforcing my conclusion on the need for a new philosophic basis.
Nevertheless, all planners need to know of modern literature in policy analysis and systems analysis, which provides methodologies directly relevant for planning. Relatively good introductions are Dunn, 1981; Brewer and deLeon, 1983; Quade, 1982; Miser and Quade, 1985. Many important relevant ideas will be included in a forthcoming book on policy analysis by Giandomenico Majone.

[3] For readable introductions, see White, 1968, and Raz, 1978. For deeper study, unequalled are the various volumes of Lenk, 1977-1981.

some of the main features of planning looked at as a mode of policy-reasoning. For the limited purposes of this essay I select some of the features which are salient to the building of bridges between abstract thoughts and concrete solutions and which fit the needs of planning as an applied "profession" which must break through its present limits on "reflection-in-action."[4]

To do so, this essay explores six main aspects of planning which characterise it as a special mode of policy-reasoning, with attention to both their theoretic and their applied implications. But, before doing so, a clarification of the concept of "planning" as used here is in order. "Planning" is a verbal symbol used in different contexts to denote quite disparate phenomena. Changing political attitudes to the term "planning," such as in the U.S.A., influence its uses and move the borderlines, if any, between it and related concepts such as "policy analysis." The absence of a verbal distinction between "policy" and "politics" in most languages and the utilisation of terms such as "political planning," in German for instance, as a substitute for "policy making" as used in English, further complicate the picture.

Therefore, semantic clarification of the meaning of the term "planning" as used in different worlds of discourse, together with typologies of "planning" and their different facets (Dror, 1971, Chapter 10), are needed for deeper treatments of planning theory and its advancement. But, for the limited purposes of the present paper, much less will do. I think that the six issues explored in the following are shared, with minor variations, by most of "planning" in its various meanings, with the exclusion of relatively simple, mainly physical projects. There may be differences in degree in the applicability of the problematics as developed in this paper to various levels of planning, ranging from "macro" and "strategic" to "micro" and "tactical."[5] But, in principle, I think that the issues examined in this essay are shared by most of planning endeavours - public and private, national and regional, physical as well as economic, defense as well as social, in highly industrialised and in very poor countries and so on. Therefore, the subject matter of this essay properly belongs to an urgently needed "general theory of planning," as a main dimension of policy-reasoning as applied to the specific features of planning.

---

[4] See Schon, 1983, especially Chapter 7, which discusses limits to reflection-in-action in town planning.

[5] Still very useful are the distinctions proposed in Anthony, 1965.

PLANNING AS DELIBERATE, "RATIONAL" AND "PROFESSIONAL" INTERVENTION WITH HISTORY

The most ambitious view of planning is to look at it as a mode of deliberate intervention with history. As compared with other modes of deliberate intervention with history, such as heroic leadership or purely intuition based statecraft, planning presumes to be "rational" and "professional." Whether planners are fully aware of it or not, planning is thus a very presumptuous activity. After all, there are few more *hubris* activities than to try and intervene with history so as to make the future better fit our desires; and to do so on the basis of what we claim is rationality and professionalism makes it all the more presumptuous.

Prescriptively, to try and intervene with history makes an understanding of historic processes essential. Once naîve notions of social change are left behind (Boudon, 1986) and economics and similar "sciences of the artificial" (Simon, 1981) are recognised for what they really are, namely time-bound and domain-bound models of the mind of some utility for managing a number of variables when others are relatively stable, the conscientious planner must strive for at least some deeper understanding of historic processes with which he wants to intervene. This, in turn, requires a good sense of historic processes in general, with all its difficulties,[6] and detailed study leading to maximal deep understanding of the particular processes at which a concrete planning project is directed.

Related is the need to diagnose a particular planning situation within historic processes. This must include, for instance:

- Locating relevant situations as on ascending or descending historic curves, as illustrated by the failure of French defense planning between World War I and World War II because of misreading of the situation and accordingly the selection of incrementally good but basically useless choices (Young, 1978), instead of the needed trend toward interventions which were the only way to break the growing German superiority.
- Identifying realistically the longer range availability of resources, so as to avoid utopian planning based on false expectations concerning resources.[7]

----

[6] Useful, as a first approach, is Neustadt and May, 1986. But, deeper understanding of historic processes is needed, as illustrated by Braudel, 1980; Faber and Meier, 1978.

- Putting specific planning projects within large national historic endeavours such as development,[8] as contrasted with the quite different planning modes appropriate for relatively isolated projects.
- Related, on a smaller scale, is the extent to which specific planning projects are part of a national policy, such as land reclamation in The Netherlands (Dutt and Costa, 1985) and industrial policies in many European countries (Katzenstein, 1985).
- Careful consideration of the needed scale of intervention, so as to achieve the critical mass required to exert desired influence on future history, on the micro as well as on the macroscale (Schulman, 1980).

Viewing planning as a mode of deliberate "rational" and professional intervention with historic processes has many additional implications, such as the need for planners to arrive at good understanding of socio-political processes and good knowledge of intervention strategies. Also important are moral implications, such as the need for appropriate codes of behaviour fitting a profession specialising in influencing the future, including the very difficult issues stemming from the very idea to try and shape the future in accordance with present values. Especially crucial for the very idea of planning is the notion of "rationality" on which it is based, to which I now turn.

----

[7] This requires overall consideration of conditions which will face policy making in the foreseeable future, as conditioning specific planning projects, see Dror, 1986, Chapter 2. Also needed is location of planning activities within historic cycles of the public activity - private concern nature. See Hirschman, 1981, and Schlesinger, 1986.
[8] For instance, all planning endeavours in Israel must be considered within the overall context of Zionism and upbuilding of Israel. See Dror, in preparation. The presence or absence of overall "national purposes" may be a main feature differentiating planning between countries and periods.

UNDERLYING NOTIONS OF "RATIONALITY"

Planning must be based on a notion of "rationality" which goes far beyond its usual meanings, as used in operations research, systems analysis, economics, etc. I therefore propose the term "ultra-rationality" for the advanced and unconventional types of "rationality" on which policy-reasoning in general and planning in particular should be based.

To clarify some of the features of ultra-rationality, let me proceed inductively by way of illustrations:

- A main underlying ultra-rationality of planning is "self-binding" (Elster, 1979 and Schelling, 1984) in the sense of limiting decision freedom in the future so as to assure somewhat against "weakness of political will"[9] and other decision spoilers.
- Basically "extra-rational" dimensions, such as creativity, fulfill essential functions in planning and must, therefore, be accounted for in ultra-rationality.[10]
- Ultra-rationality as a base for planning must accommodate paradoxical situations where self-delusion may be useful, for instance in the face of extreme adversity when only self-fulfilling prophecy effects may succeed despite overwhelming odds. As a result, sophisticated planning must recognise conditions where planning is an inappropriate mode of policy-reasoning and conditions when quite different decision modalities, such as visionary leadership, may be preferable (Dror, 1986, pp. 95-97).
- Irrationality must be fully taken into account in ultra-rationality, both as a widespread fact of human behaviour,[11] and sometimes,

----

[9] The issue of *akrasia*, e.g. weakness of will, well illustrates the need to develop a philosophy of policy-reasoning, as concepts taken from action on the level of individuals, however suggestive, often do not fit collective action without much reworking. For example, see the treatment of "weakness of will" in Davidson, 1980.

[10] The "optimal model" presented in Dror, 1983, while needing further development to fit the requirements of ultra-rationality, discusses at length the importance of "extra-rational" components of a "rational" basis for policy making including planning.

[11] The neglect of mass psychology in most of North American planning thinking reflects its disregard in Anglo-American social sciences in general. This leads to interesting issues of the dependance of planning, as an intellectual endeavour and a disciplinary venture,

as a preferable posture based on the idea of "rationality of irrationality."[12]

- Paradoxes and antinomies, both as features of social reality and as essential for handling complex issues, must be included in ultra-rationality (Hofstadter, 1979, 1985). These include, for instance, the very important relations between micromotives and macrobehaviour (Schelling, 1978) and a variety of "paradoxes of politics" (Brams, 1976). This point deserves emphasis as a counterbalance to the very simplistic nature of many of the notions of rationality posed as a basis for planning (Churchman, 1979, Elster, 1978).
- Ultra-rationality must handle the self-contradictory task of improving value judgment, which basically is an extra-rational process, but one crucial for planning. What is needed, in essence, are ways for "rationally" upgrading an "extra-rational" process, for instance by structuring the field for value judgment without interfering with the judgment itself. The need to choose between absolute values well illustrates the problem, which is an important challenge to ultra-rationality (Bobbitt and Calabrasi, 1979, Levi, 1986).
- Ultra-rationality must overcome the "individual-collective" dichotomy in an integrative way, as an essential need for handling realities effectively (Giddens, 1984, Coleman, 1986).

These are only some of the facets and issues of ultra-rationality. Construction of models of ultra-rationality as a basis for planning is an urgent task which depends on advancement of a philosophic foundation for policy-reasoning. Within the confines of this essay, only one additional feature of ultra-rationality can be examined, namely "debugging" as a central mode of decision improvement and, therefore, as a main rationale of planning.

----

on the biases of the cultures in which it is located. To counteract this particular neglect, some study of mass psychology is recommended to all planners (Moscovici, 1986).

[12] The idea of "rationality of irrationality" has been developed in strategic thinking and planning (Schelling, 1984, Kahneman *et al.*, 1982, Dror, 1980).

"DEBUGGING" AS A MAIN RATIONALE

One additional strategy of ultra-rationality is of pivotal importance to planning, though shared with other modes of policy-reasoning, namely "debugging." The very complexity of ultra-rationality and its multidimensionality precludes the idea of arriving at any simple model which can serve as an ideal for planning. Therefore, a reverse strategy is all the more important, namely a striving for reduction of decision errors - called by me, adopting computer programming language, "debugging." Here, planning is regarded as a way to improve policy-reasoning by exposing and offsetting the main prevalent policy-reasoning fallacies and mistakes.

Strictly speaking, identification of decision errors also depends on the existence of some notions of nonerroneous decision making. But such notions are much less demanding than explicated and comprehensive models of ultra-rationality. Therefore, the debugging strategy is of much practical importance and deserves emphasis in efforts to build up a more advanced planning theory and practice.

This important strategy of planning as a mode of policy-reasoning is again best presented by a number of illustrations. To diversify somewhat the mode of presentation, I will in this case present the illustrations in the form of positive recommendations, directed at counteracting widespread actual decision-making weaknesses (Elster, 1983, 1986, Hogwood and Peters, 1985, Janis, 1982, Janis and Mann, 1977, Kahneman, Slovic and Tversky, 1982, Pears, 1984):

- Planning should be characterised by "cold" reasoning, to counterbalance the widespread tendency towards "red-hot" and even "white-hot" decision making, especially under conditions of pressure and in the wake of calamities.[13]

----

[13] This raises some interesting problems: one of the functions of planning is to prepare contingency plans for crises. Also, planning should prepare good ideas, even if nonfeasible, for possible "windows of opportunity," such as following calamities. But, planning itself is not immune to "hot thinking" and rushing into projects under the impact of catastrophies. The Delta words in The Netherlands illustrate such a case combining utilisation of a passing opportunity and some elements of unjustified "hot" decision making. See Dutt and Costa, 1985, especially Chapter 12.

- Planning should be characterised by "deep" thinking, as contrasted to the surface, if not superficial, thinking of most of actual decision making.
- Planning should follow an explicit protocol for handling uncertainty, to overcome the widespread fallacies of human thinking in the face of uncertainty.
- Planning should take a coherent view of larger systems of issues and problems, as contrasted with the piecemeal handling of issue fragments universal in governments and organisations.
- Planning should be issue oriented, as a counterbalance to the micropolitical considerations dominating much of policy making.

The above list can and should be much extended. But, it serves to illustrate the idea of "debugging" as a main rationale of planning. It also indicates an additional important role of planning, related to the debugging function: planning should serve as a tutorial, educating decision makers to comprehend more correctly issues and select better options, and indeed to think less erroneously on complex predicaments.[14]

SYNCHRONIC AND DIACHRONIC COHERENCE

Debugging is an important strategy of planning but not unique to it, being shared with other modes of policy-reasoning. In contrast, synchronic and diachronic coherence is of the essence of planning and constitutes its single most important unique characteristic. Thus, policy analysis and systems analysis in most of their versions, despite disclaimers to the contrary, tend to focus on discrete decisions.

The synchronic and diachronic coherence feature is well recognised in planning literature and practice, so I can here deal with it very briefly, limiting myself to three main observations: first, coherence does not imply the extreme of "comprehensive" planning and does not exclude strategies of shock treatments and concentration

---

[14] The view of planning as education of decision makers leads to additional vistas of planning related to its organisational dimensions and its interfaces with organisational, national and societal choice processes. It raises even more vexing issues, such as the tension between the idea of planning as societal critique and iconoclasm on one hand and the roles of planning as a "servant of power" on the other hand.

on leading sectors or projects. But, in the latter cases too, planning should adopt a broad time and domain horizon for consideration and deal with a coherent cluster of actions directed at having impact on the future.

Second, the synchronic and diachronic coherence requirement raises many problems, such as preferable scope, appropriate time scales, needed interdisciplinary knowledge, etc. However recognised and despite receiving a lot of attention, these issues are far from adequately handled on the level of planning theory and need innovative work.

Third, the synchronic and diachronic coherence requirement depends on capacities to cope with complexity. Despite some progress (Baldwin, 1975, La Porte, 1975, Warfield, 1976, Dror, 1986, pp. 165-67), no satisfactory theoretic basis for handling complexity is available. In particular, nonlinear and noncontinuous relations between variables pose challenges to policy-reasoning and adequate solutions for these problems are essential for planning.

A striking expression of the weaknesses of the present state of complexity treatment in planning is mishandling of uncertainty, which is at the core of striving for coherence. Uncertainty is a main limit on synchronic and diachronic coherence and better handling of uncertainty is a *sine qua non* for upgrading planning. I think that the issue of uncertainty requires quite a radical break in much of our planning theory and practice, based on a novel conception of planning as a "fuzzy gambling" activity. This is a major proposed revision in planning theory with many practical implications, and deserves closer examination within this paper.

PLANNING AS "FUZZY GAMBLING"

The view of planning as "fuzzy gambling" is radical in its contents and implications, even though the facts of uncertainty and the frequency of unanticipated consequences as a main bane of planning (Boudon, 1982, Sieber, 1981) are well known. Much of my present work concentrates on this subject,[15] which I am far from having

----

[15] Because of its relevance for the relations between abstract thinking and practice, may I mention that my "planning as fuzzy gambling" insight and then conjecture resulted directly from practical work as Senior Policy Analysis and Planning Advisor of the Israeli Ministry of Defense.

worked out to my own satisfaction.[16] The most convenient way to proceed is to move through a number of propositions, which add up to the conclusion that part of the core essence of planning is "fuzzy gambling."

1. *All planning is faced by many uncertainties.*

This statement is so self-evident that it scarcely needs elaboration, the only exception being very narrow cases of microplanning, mainly of simple physical projects. Less obvious and very important is the tendency of uncertainty to expand, because of the increasingly dynamic and partly jumpy nature of many of the variables relevant to planning, ranging from technological innovations and their implications to political culture and social values; and including the demographic infrastructure, as well as economic resources, not to speak about war and peace.

2. *Many of the uncertainties facing planning are both "hard" and far-reaching.*

Many of the uncertainties facing planning are, according to present world views, inherent in the dynamics of relevant processes - which are stochastic and often "arbitrary" and indeterminate in nature. By definition, such "hard" uncertainties are not a consequence of limited prediction capacities of human knowledge, including prediction methodologies, but persist even in the face of hypothetically "perfect planners" who know all facts and their true dynamics.

Not only are many uncertainties facing planning "hard," but they are far-reaching, in the sense of relevant processes often being "explosive" rather than smooth, involving ultra-change (i.e. change in the patterns of change itself); and including uncertainty concerning the very shape of possible futures and not only their probability distributions.[17]

----

[16] For a preliminary formulation, see Dror, 1986, especially pp. 97-98, 167-76. For application to planning, see my paper "Planning as Fuzzy Gambling: A Radical Perspective on Coping with Uncertainty," in Morley and Schachar (eds.), 1986, pp. 24-39. I am now working on a book on *Policy-Gambling* which hopefully will work out this idea and its theoretic as well as operational implications, also for planning.

[17] The appropriate mathematical models are catastrophe theory and

3. *As a consequence, many of the uncertainties facing planning cannot be reduced by any prediction method and should not be handled with "subjective probabilities."*

Hard uncertainties are inherent in relevant processes. Therefore, even perfect prediction methods, which are far from being available, cannot reduce them. Instead, perfect prediction can map hard uncertainties and declare on them that "we know that we cannot know" - which is a radically different statement than "I do not know." It follows, that subjective probabilities, as advocated in most proposals for handling uncertainty in planning,[18] have no sound epistemological justification when hard uncertainties are faced: they produce misleading illusions and not helpful approximations of the probability of future possibilities.[19]

4. *Decisions in the face of hard and far-reaching uncertainties constitute "fuzzy gambles."*

However unpleasant, the unavoidable conclusion is that decisions must be regarded as "fuzzy gambles," that is gambles with undefined and in part indeterminate payoff functions and "rules of the game." The more a decision faces harder and far-reaching uncertainties, the more it constitutes a "fuzzy gamble," with all that this implies.

----

even chaos mathematics rather than smooth curves, which still dominate textbooks. For an introduction, see Woodcock and Davis, 1978. Professional planners should absorb more advanced versions, at least as essential metaphors for perceiving and handling reality, such as presented in Thom, 1983.

[18] Thus, incorrect despite its pioneering contributions is, in my view, part of the classic book by Mack, 1971. Subjective probabilities continue to dominate professional literature on coping with uncertainty.

[19] The philosophic and epistemological basis of probability is essential knowledge for all planners. It serves as a basis for handling the uncertainties inherent in planning. An easy historic introduction is Hacking, 1975. More advanced planning relevant treatments are represented by Suppes, 1984.

5. *It follows that planning is a fuzzy gambling activity and/or a fuzzy gambling aid.*

Planning is distinguished by its longer time horizons and by its effort to deal with interacting sets of variables, in contemporary as well as time axes. These features of planning augment the hardness and extremity of uncertainties with which it must cope, making it - depending on whether a form of planning is more of a decision-making or a decision-advising activity - a fuzzy gambling activity and/or a fuzzy gambling aid.

Taking the five propositions together, this "fuzzy gambling" view of planning has far-reaching consequences for the external outputs of planning as well as for required changes in the internal features of planning. Some of the external consequences include:

- Unexpected and often undesired consequences are inherent to planning and cannot be avoided even in hypothetical perfect planning.
- To push the above observation to one of its logical conclusions, the quality of planning cannot be validly judged by its fit with future conditions. When one "gambles," results depend in part on "chance," so that often very good planning may fail and sometimes very bad planning may succeed. Usually clients and publics will not accept such logic, insisting on holding planning accountable in terms of results. This is politically unavoidable, even if intellectually unjustified.
- Even aside from the above problem, planning will disappoint many expectations by political clients, who often hope that planning will reduce uncertainties. Therefore, it may well be necessary to educate top decision makers to the realities of policy-reasoning and especially planning as "fuzzy gambling," to permit correct expectations from planning and its appropriate utilisation.
- In the absence of policy-gambling, sophisticated clients and publics, in risk-adverse contemporary cultures (Douglas and Wildavsky, 1982, Douglas, 1986) planning faces the difficult moral issue of how far to explicate its "fuzzy gambling" nature.

Pertinent for our purposes are the internal implications, which are radical in requiring far-reaching retrofitting of planning approaches and methods and, even more difficult, fundamental changes in the self-perception of planners. Let me illustrate this point:

- As a first step, a more strenuous effort must be made to reduce uncertainties as much as validly possible, by full but correct utilisation of all available prediction methods. This means that prediction methodology must become an integral part of the equipment of planning (Helmer, 1983, Martino, 1983).
- Most of the present methods for handling uncertainty in planning must be used much more carefully and, in part, abandoned. Instead, a whole array of different methods must be adopted, ranging from much more social experimentation (Ferber and Hirsch, 1982) to "adaptive management,"[20] with realistic risk analysis[21] being an essential dimension in any planning activity.
- The very idea of a "plan" needs revision, as recognised in some of the better planning practice. Instead, "guidelines," "principles," "planning frames," etc. are often preferable. To be sure, detailed multiyear plans are archaic. This, in turn, aggravates the difficult relationships between planning on one side, and implementation and budgeting on the other side, which need reconsideration.
- The nature of planning as a continuous process, based on constant adjustments to emerging realities and to feedback needs strengthening. This ties in to looking at planning as an important part of societal and governmental learning processes, which deserve more attention.[22]
- The dual relationships with uncertainty, with planning, on one hand, adjusting itself to irreducible uncertainties, and planning, on the other hand, trying to shape future realities and thus, in part, reduce uncertainties, must be recognised with appropriate consequences for planning. For instance, the already mentioned need for critical intervention masses is reinforced when order must be imposed on uncertainties.

----

[20] Very relevant and important is Holling, 1978. Other important approaches are illustrated by Ascher and Overholt, 1983.

[21] A good introduction is Baruch *et al.*, 1981. Somewhat different approaches are illustrated by Schwing and Albers (eds.), 1980. Advanced planning relevant work is illustrated by Miller, Kleindorfer and Munn, 1986.

[22] See Michael, 1973, to be compared with studies of actual governmental nonlearning, e.g. Etheredge, 1985. Here, some engineering thinking may be very relevant for planning (e.g. Petroski, 1985). Possible functions of planning as an aid to organisational learning are illustrated by Mason and Mitroff, 1981.

- Building up resilience is a main goal of planning in the face of uncertainty, instead of widespread demands for "risk avoidance" (Wildavsky, forthcoming).
- The debugging functions of planning as fuzzy gambling are especially pronounced, because of the widespread actual fallacies of decision making under uncertainty.
- The role of planning as education of decision makers too is very pronounced in respect to uncertainty, adequate understanding of the nature of policy-gambling and the roles of planning in it being essential, as already mentioned.
- The value dimension of planning under uncertainty must be faced systematically, involving usually repressed "lottery value" judgment on preferences between different bundles of risks and uncertainties (Raiffa, 1968).
- Planners must become uncertainty sophisticated, on the cognitive as well as the intuitive level.[23] And the idea of planning as fuzzy gambling must be integrated into the self-image of planning.
- Planners must be aware of the limits of planning when uncertainties become extreme. Under such conditions, crisis management and improvisation become preferable modes of policy-reasoning with far-reaching revisions of the roles of planning.[24]
- Planning organisations themselves must become adaptive and very dynamic in all respects (de Greene, 1982).

These observations are but a cursory introduction to the subject of planning as a fuzzy gambling activity and to its many implications, which should be in the centre of the work program of the advancement of planning theory and practice.

---

[23] However strange this may appear, I recommend to planners to read carefully, at least as a metaphor, Hayano, 1982.

[24] About twenty years ago, I struggled with the riddle of the tremendous successes of Israel despite absence of planning in any meaningful sense (Akzin and Dror, 1966). Today I see the answer clearly. Under conditions of far-reaching uncertainty, as in Israel, good improvisation is far superior to traditional planning, which assumes much predictability. Sophisticated planning was not available then and is hardly available even now.

ANALYSIS AND INNOVATIVENESS

To wind up the perspectives on planning as a policy-reasoning mode presented in this essay, it is appropriate to conclude with the special feature of planning as including both analysis and innovativeness, with all the internal tensions this implies.

In principle, one can distinguish in policy-reasoning two main processes, which co-exist with difficulty: options must be generated, in large part by invention and other innovation processes; and options must be screened, to select the relatively best one to be adopted.[25] The trouble is, that these two processes are different in their bases, option generation being an extra-rational and partly "wild" process, while option screening involves rational, often rather narrow, "cost-benefit" analytical processes.

Leaving aside the interesting question whether the difference between these processes has a neurological basis in division of functions between the two hemispheres of the brain, in terms of personal aptitudes, organisational climates, appropriate training and development of human capacities and degree of possible reliance on explicated methods - there is a world of difference between the creative, option innovating, process, and the screening between available options process. As a matter of fact, most decision improvement approaches, such as policy analysis, neglect the option innovation process and concentrate on the methodologically "harder" option screening process. But this is a serious omission, because no screening can arrive at a better option than the best of those considered, perhaps incrementally improved as a result of the screening. This weakness is all the graver because changes in predicaments, as characterises much of present and emerging situations, require new option invention as most of the ones given in the past are inadequate. In other words, option innovation, including much option invention, is a main requisite of good policy-reasoning.

This requirement makes planning all the more important a mode of policy-reasoning, because planning emphasises the importance of "design" in the sense of invention and development of new options.

---

[25] The distinction goes back to Popper, 1963. A relevant discussion and development of Popper is Albert, 1978, and Albert, 1980. More recent semi-evolutionary approaches to the advancement of knowledge, stimulating to apply to planning in particular and policy-reasoning in general are illustrated by Munz, 1985, and Wuketits, 1984.

Therefore, planning can meet a growing need of policy-reasoning, on the condition that its creative parts are pushed ahead without neglecting the screening needs. Accordingly, the dual nature of planning, seen as including innovativeness and screening and trying to combine them, deserves strengthening and advancement.

Indeed, it may be worthwhile to go one step further and take a fresh look at planning as, in part, an entrepreneurial process. Parts of the failures of "strategic planning" in business enterprises can be explained in terms of lack of attention to the innovation dimensions of planning (Steiner, 1979). What is needed instead is a restructuring of planning, both public and private, to meet the needs of more entrepreneurship - to be served by appropriately revised planning.[26]

SOME IMPLICATIONS FOR PRAXIS

The above preliminary exploration of six main issues of policy-reasoning as applied to planning leaves many additional problems and approaches for more extensive effort.[27] But it seems to me that a basis has been provided which justifies at least three preliminary recommendations bridging thinking and practice:

----

[26] Therefore, concepts of "entrepreneurship" are relevant to advancement of planning theory and practice. The relationships between planning and growing needs for entrepreneurship need to be worked out. Relevant introductory readings include Lewis, 1980, to be read together with Meritt and Merritt (eds.), 1985, on one hand; and Polsby, 1984, on the other hand. More economics oriented approaches are illustrated by Casson, 1982.
Planning theory and practice must also redefine its relations to business corporation needs and problems, as discussed for instance, in Peters and Waterman, 1982; and further developed, in different but still relevant for planning directions, in Peters and Austin, 1985. The latter book also raises the problems of "planning" and "innovative leadership," which add an additional important dimension to the question, *Planning Quo Vadis*?

[27] For example, an additional approach to planning is to view it as a main dimension of a general societal control and direction theory. For some relevant ideas, see for instance Geyer and Van der Zouwen (eds.), 1986.

1. *In-depth work on planning theory is needed, in conjunction with the advancement of other policy-oriented disciplines and on the basis of development of a philosophic basis for policy-reasoning.*
2. *Planning professionals must reconsider many of their basic assumptions and revise many of their practical tools. Indeed, the self-image of planning as a profession requires a fresh look.*
3. *Radical changes are needed in the professional preparation of planners, including new types of planning schools and crash programs to "update" present planners.*

These recommendations add up to a very demanding view of the functions of planners and on the qualities they need.[28] This, I think, is justified by broad social and human needs in the face of foreseeable problematics: states urgently need vastly augmented capacities to handle democratically very complex predicaments, while the penalties for failure add up to endangerment of the very survival of humanity. To build up required democratic capacity to govern (and, *mutatis mutandis*, parallel capacities on the level of private enterprises, intermediate bodies, voluntary associations, etc.), one of the essential needs is for an improved professional cadre of planners, to provide pivotal inputs into decision making.

Traditional planning theory and practice is a main basis for developing the needed new planning cadres. But, in order to move ahead, additional perspectives on planning are needed based on a broader conception, and also philosophy, of policy-reasoning. This essay presents some thoughts in this direction, inspired inter alia by the pioneering example of my friend from Harvard study days, Peter Nash.

---

[28] My demanding conception of planners as a crucial cadre in societal attempts to influence the future leads, at the very least, and as a first step, to the conclusion that all who want to be full fledged professional planners in the macrosense of my essay (as distinct from "planning technicians"), hopefully including the readers of this essay as well as myself - are morally committed to regard "planning" as their life mission and life project. This implies, in addition to many ethical, career pattern and other implications, that we must engage in continuous and very strenuous learning. The books cited in the references are suggested as initial readings in such an endeavour.

REFERENCES

Akzin, B. and Dror, Y. (1966) *Israel: High-Pressure Planning*, Syracuse, New York: Syracuse University Press.

Albert, H. (1978) *Trakat ueber Rationale Praxis*, Tuebingen: J.C.B.Mohr.

Albert, H. (1980) *Trakat ueber Kritsche Vernunft*, Tuebingen: J.C.B. Mohr.

Ascher, W. and Overholt, W.H. (1983) *Strategic Planning and Forecasting: Political Risk and Economic Opportunity*, New York: Wiley.

Anthony, R.N. (1965) *Planning and Control Systems: A Framework for Analysis*, Boston: Harvard University, Graduate School of Business Administration.

Baldwin, M.M. (ed.) (1975) *Portraits of Complexity: Application of Systems Methodologies to Societal Problems*, Columbus, Ohio: Battelle Memorial Institute.

Baruch, F. *et al.* (1981) *Acceptable Risk*, Cambridge: Cambridge University Press.

Boudon, R. (1982) *The Unintended Consequence of Social Action*, London: Macmillan.

Boudon, R. (1986) *Theories of Social Change: A Critical Appraisal*, Cambridge, U.K.: Polity Press.

Bobbitt, P. and Calabrasi, G. (1979) *Tragic Choice*, New York: Norton.

Brams, S.J. (1976) *Paradoxes in Politics: An Introduction to the Nonobvious in Political Science*, New York: Free Press.

Braudel, F. (1980) *On History*, Chicago, Ilinois: University of Chicago Press.

Brewer, G.D. and deLeon, P. (1983) *The Foundations of Policy Analysis*, Homewood, Illinois: Dorsey, 1983.

Casson, M. (1982) *The Entrepreneur: An Economic Theory*, Totowa, New Jersey: Barnes and Noble.

Churchman, C.W. (1979) *The Systems Approach and Its Enemies*, New York: Basic Books.

Coleman, J.S. (1986) *Individual Interests and Collective Action: Selected Essays*, Cambridge, U.K.: Cambridge University Press.

Davidson, D. (1980) *Essays on Actions and Events*, Oxford: Oxford University Press.

de Greene, K. (1982) *The Adaptive Organisation: Anticipation and Management of Crisis*, New York: Wiley.

Douglas, M. (1986) *Risk Acceptability According to the Social Sciences*, London: Routledge and Kegan Paul.

Douglas, M. and Wildavsky, A. (1982) *Risk and Culture*, Berkley, California: University of California Press.

Dror, Y. (1971) *Ventures in Policy Sciences: Concepts and Applications*, New York: American Elsevier, ch. 10.

Dror, Y. (1980) *Crazy States: A Counterconventional Strategic Problem*, (supplemented ed.), Millwood, New York: Kraus Reprints.

Dror, Y. (1986) *Policy Making Under Adversity*, New Brunswick: New Jersey: Transaction, 162-65.

Dror, Y. (19) *Breakthrough Policies for Israel: Memorandum for the Prime Minister*, (in preparation).

Dunn, W.N. (1981) *Public Policy Analysis: An Introduction*, Englewood Cliffs, New Jersey: Prentice-Hall.

Dutt, A.K. and Costa, F. (eds.) (1985) *Public Planning in the Netherlands*, Oxford: Oxford University Press.

Elster, J. (1979) *Ulysses and the Sirens: Studies in Rationality and Irrationality*, Cambridge: U.K.: Cambridge University Press.

Elster, J. (1978) *Logic and Society: Contradictions and Possible Worlds*, New York: Wiley.

Elster, J. (1983) *Sour Grapes: Studies in the Subversion of Rationality*, Cambridge, U.K.: Cambridge University Press.

Elster, J. (1986) *The Multiple Self*, Cambridge, U.K.: Cambridge University Press.

Etheredge, L.S. (1985) *Can Governments Learn? American Foreign Policy and Central American Revolutions*, New York: Pergamon.

Faber, K.G. and Meier, C. (1978) *Historische Prozesse*, Munich: Deutscher Taschenbuch Verlag.

Faludi, A. (1973) *Planning Theory*, Oxford: Pergamon Press.

Faludi, A. (1986) *Critical Rationalism and Planning Methodology*, London: Pion.

Ferber, R. and Hirsch, W.Z. (1982) *Social Experimentation and Economic Policy*, Cambridge, U.K.: Cambridge University Press.

Geyer, F. and Van der Zouwen, J. (eds.) (1986) *Sociocybernetic Paradoxes: Observation, Control and Evolution of Self-steering Systems*, London: Sage.

Giddens, A. (1984) *The Constitution of Society: Outline of the Theory of Structuration*, Cambridge: Polity Press.

Graumann, C.F. and Moscovici, S. (eds.) (1986) *Changing Conceptions of Crowd Mind and Behaviour*, New York: Springer-Verlag.

Hacking, I. (1975) *The Emergence of Probability*, Cambridge, U.K.: Cambridge University Press.

Hayano, D.M. (1982) *Poker Faces: The Life and Work of Professional Card Players*, Berkley, California: University of California Press.

Healey, P. McDougall, G. and Thomas, M. (eds.) (1982) *Planning Theory: Prospects for the 1980s*, Oxford: Pergamon Press.

Hîrschman, A.O. (1981) *Shifting Involvement: Private Interest and Public Action*, Oxford: M. Robertson.

Hofstadter, D.R. (1979) *Goedel, Escher, Bach: An Eternal Golden Braid*, New York: Basic Books.

Hofstadter, D.R. (1985) *Metamagical Themas: Questing for the Essence of Mind and Pattern*, New York: Basic Books.

Hogwood, B.W. and Peters, B.G. (1985) *The Pathology of Public Policy*, Oxford: Claredon Press.

Holling, C.S. (ed.) (1978) *Adaptive Environmental Assessment and Management*, New York: Wiley.

Janis, I.L. (1977) *Decision Making: A Psychological Analysis of Conflict, Choice, and Commitment*, New York: Free Press.

Janis, I.L. (1982) *Groupthink*, (2nd. ed.), Boston: Houghton Mifflin.

Jennergren, C.G. (ed.) (1978) *Trends in Planning*, Stockholm: Swedish National Defense Research Institute.

Kahneman, D., Slovic P. and Tversky, A. (1982) *Judgment Under Uncertainty: Heuristics and Biases*, Cambridge: Cambridge University Press.

Katzenstein, P.J. (1985) *Small States in World Markets: Industrial Policy in Europe*, Ithaca, New York: Cornell University Press.

La Porte, T.R. (ed.) (1975) *Organised Social Complexity: Challenge to Politics and Policy*, Princeton, New Jersey: Princeton University Press.

Lenk, H. (ed.) (1977-1981) *Handlungstheorien Interdisziplinaer*, Munich: Wilhelm Fink.

Levi, I. (1986) *Hard Choices: Decision Making Under Unresolved Conflict*, Cambridge, U.K.: Cambridge University Pres.

Lewis, E. (1980) *Public Entrepreneurship: Towards a Theory of Bureaucratic Political Power - The Organisational Lives of Hyman Rickover, J. Edgar Hoover and Robert Moses*, Bloomington: Indiana University Press.

Mack, R.P. (1971) *Planning on Uncertainty: Decision Making in Business and Government Administration*, New York: Wiley.

Mack, R.P. and Martino, J.P. (1983) *Technological Forecasting for Decision Making*, (2nd. ed.), New York: Elsevier.

Mason, R.D. and Mitroff, I.I. (1981) *Challenging Strategic Planning Assumptions: Theory, Cases and Techniques*, New York: Wiley.

Meritt, R.L. and Merritt, A. (eds.) (1985) *Innovation in the Public Sector*, Beverly Hills, California: Sage.

Michael, D.M. (1973) *On Learning to Plan - and Planning to Learn: The Social Psychology of Changing Toward Future Responsive Societal Learning*, San Francisco: Jossey-Bass.

Miller, C.T., Kleindorfer, P.R. and Munn, R.E. (1986) *Conceptual Trends and Implications for Risk Research: Report of the Task Force Meeting "Risk and Policy Analysis Under Conditions of Uncertainty,"* Laxenburg, Austria: International Institute for Applied Systems Analysis, Report CP-86-26.

Miser, H.J. and Quade, E.S. (eds.) (1985) *Handbook of Systems Analysis: Overview of Uses, Procedures, Applications, and Practice*, New York: North-Holland.

Morley, D. and Schachar, A. (eds.) (1986) *Planning in Turbulence*, Jerusalem: Magnes Press.

Moscovici, S. (1985) *The Age of the Crowd: A Historical Treatise on Mass Psychology*, Cambridge: Cambridge University Press.

Munz, P. (1985) *Our Knowledge of the Growth of Knowledge: Popper or Wittgenstein?* London: Routledge and Kegan Paul.

Neustadt, R.E. and May, E.R. (1986) *Thinking in Time: The Uses of History for Decision Makers*, New York: Free Press.

Pears, D. (1984) *Motivated Irrationality*, Oxford: Claredon Press.

Peters, T.J. and Waterman, R.H. Jr. (1982) *In Search of Excellence: Lessons from America's Best Run Corporation*, New York: Harper and Row.

Peters, T. and Austin, N. (1985) *A Passion for Excellence: The Leadership Difference*, New York: Random House.

Petroski, H. (1985) *To Engineer is Human: The Role of Failure in Successful Design*, New York: St. Martin's Press.

Polsby, N.W. (1984) *Political Innovation in America: The Politics of Policy Initiation*, New Haven, Colorado: Yale University Press.

Popper, K.R. (1963) *Conjectures and Refutations: The Growth of Scientific Knowledge*, London: Routledge and Kegan Paul.

Quade, E.S. (1982) *Analysis for Public Decisions*, (2nd. ed.), New York: North-Holland.

Raiffa, H. (1968) *Decision Analysis*, Reading, Massachusetts: Addison-Wesley.

Raz, S. (ed.) (1978) *Practical Reasoning*, Oxford: Oxford University Press.

Schelling, T.C. (1978) *Micromotives and Macrobehaviour*, New York: W.W. Norton.
Schelling, T.C. (1984) *Choice and Consequences: Perspectives of an Errant Economist*, Cambridge, MA: Harvard University Press.
Schlesinger, A.M. Jr. (1986) *Cycles of American History*, Boston: Houghton Mifflin.
Schon, D.A. (1983) *The Reflective Practitioner: How Professionals Think in Action*, New York: Basic Books.
Schulman, P.R. (1980) *Large Scale Policy Making*, New York: American Elsevier.
Schwing, R.C. and Albers, W.A. (eds.) (1980) *Societal Risk Assessment: How Safe is Safe Enough*, New York: Plenum.
Sieber, S.D. (1981) *Fatal Remedies: The Ironies of Social Intervention*, New York: Plenum.
Simon, H.A. (1981) *The Sciences of the Artificial*, (rev. ed.), Cambridge: MIT Press.
Steiner, G.A. (1979) *Strategic Planning*, New York: Free Press, 1979.
Suppes, P. (1984) *Probabilistic Metaphysics*, Oxford: Basil Blackwell.
Thom, R. (1983) *Mathematical Models of Morphogenesis*, New York: Halsted Press.
Warfield, J.N. (1976) *Societal Systems: Planning, Policy and Complexity*, New York: Wiley.
White, A.R. (ed.) (1968) *The Philosophy of Action*, Oxford: Oxford University Press.
Wildavsky, A. (Forthcoming) *Searching for Safety*.
Wilson, D.E. (1979) *National Planning in the United States: An Annotated Bibliography*, Boulder, Colorado: Westview.
Wilson, D.E. (1980) *The National Planning Idea in U.S. Public Policy: Five Alternative Approaches*, Boulder, Colorado: Westview.
Woodcock, A. and Davis, M. (1978) *Catastrophe Theory*, New York: E.P. Dutton
Wuketits, F.M. (ed.) (1984) *Concepts and Approaches in Evolutionary Epistemology: Towards an Evolutionary Theory of Knowledge*, Dordrecht: D. Reidel.

CHAPTER 3

PLANNING THE INFORMATION-AGE METROPOLIS: THE CASE OF SINGAPORE

Kenneth E. Corey
Department of Geography
University of Maryland
College Park, Maryland

INTRODUCTION

As one of Peter Nash's students and colleagues, this piece, for me, represents an outgrowth of the teachings, writings and career of Peter Nash. He is a futurist, a planner, a geographer, a theorist and a pragmatist. The case study and the observations that follow reflect both this combination of activities and the kind of abstract thoughts and concrete solutions given to us by Professor Nash over the years. This essay is a peek into the future. It examines how one metropolitan-scale area has planned itself for and into the information age.[1] Singapore can provide us with insight and lessons into the issues faced by a metropolis planning for an information economy, society and spatial structure.

----

[1] This is a preliminary report on research into the policies, programs and planning process used to plan an information-age metropolis. Using the city-state of Singapore as the case study, the overall research was designed to understand the nature of the proposals and interventions that have gone into that city's recent internal development.

Background on Singapore

Singapore is a city-state of 2.6 million people. It is located just north of the equator at the southern tip of the Malaysian peninsula. The Island of Singapore (570 square kilometres) and numerous other nearby tiny islands combine for a total land area of 620.5 square kilometres. The dimensions of the main island are 41.8 kilometres along the east-west axis and 22.9 kilometres on the north-south axis (Information Division, 1986, p. 1). Other than strategic location, Singapore has little in the way of natural resources. Its location and harbour long have enabled it to be one of the world's most important ports. As such, entrepot trade has dominated Singapore's economy.

In 1819, a trading station was founded on Singapore Island by Sir Stamford Raffles. Physical planning was used in those early days to lay out the various precincts of the port city. As a base for the British East India Company and later as a Crown Colony, Singapore formed a key link in the network of the British Empire. From 1942 through most of 1945, the forces of the Japanese Empire occupied Singapore. Singapore then went through various phases toward self-government until it was realised in 1959. In that year the People's Action Party (PAP) won a majority of seats to the Legislative Assembly and formed the government. The PAP has been returned to power in each of the seven general elections held through December 1984 (Information Division, 1986, pp. 13-17).

Singapore's planning and economic development are survival oriented. The small size both of Singapore's land area and its domestic market, along with its vulnerability to the swings of the world economy, plus being surrounded by neighbours with histories of tensions and hostilities towards the Republic, all seem to have combined to produce a rugged, hard working society and polity. By anticipating problems and then planning accordingly, Singapore's government, in its early years, kept its citizens several steps ahead of survival. Singaporeans project the image of "a disciplined and rational people" (Goh, 1972, p. 265). As a result, "Singapore is one of the most planning-conscious countries in the world" (Teh, 1979, p. 12).

The primary time period of concern here is with Singapore since it achieved self-government. In 1958 the United Kingdom passed legislation enabling the creation of the State of Singapore. In 1959 Singapore achieved internal self-government. Malaysia was formed in 1963, and an independent Singapore was included within the Federation of Malaysia. By 1965 Singapore and Malaysia separated, and on 9 August 1965 Singapore became a fully independent and sovereign country.

Singapore as a Case in Information-Age Planning

Singapore was selected for this study because planning has been central to its development; because it is a relatively small, clearly definable physical and statistical metropolitan scale entity; and because its economy long has been dominated by services. Most importantly, Singapore is virtually a city state.

> To understand Singapore it is important to recall that it was founded as a city, and from the very beginning has thought of itself as nothing else than a city. (Koenigsberger, 1964, p. 307)

These attributes suggested that Singapore might serve as an appropriate case for the advancement of our knowledge of urban planning in an era increasingly dominated by employment in service in general, and in information and knowledge occupations in particular.

Singapore as an Information-Age City

A recent study by Kuo and Chen (1985) revealed that Singapore has had a *service economy* since the 1920s. Its history as a major port with significant entrepot trade activities has produced a work force with a service sector of 60.7 per cent in 1980. As a result of post-colonial era planned initiatives to industrialise, secondary-sector employment in Singapore has increased from 20.9 per cent in 1957 to 37.7 per cent in 1980.

The 1980 *occupational structure* of Singapore shows that white-collar occupations comprise 41.5 per cent of total employment of 1,077,090. Blue-collar occupations rank next at 40.4 per cent; service occupations rank third at 10.4 per cent; agricultural workers and fishermen account for 1.9 per cent of the occupations; and "unclassifiable occupations" for 5.8 per cent (Kuo and Chen, 1985, p. 21).

Using the 1981 typology of information occupations by the Organisation for Economic Cooperation and Development (OECD), Kuo and Chen also found that employment in Singapore's *information occupations* in 1980 had risen to 34 per cent from 18.6 per cent in 1947 and 21.0 per cent in 1957 just before self-government (Kuo and Chen, 1985, pp. 9 and 22). The distribution among Singapore's information-worker OECD categories for 1980 was: (1) information producers, 8.1 per cent; (2) information processors, 20.2 per cent; (3) information distributors, 3.1 per cent; and (4) information infrastructure, 2.6 per cent (Kuo and Chen, 1985, p. 22). Again, these information workers represented 34 per cent of the total Singapore work force of 1,077,090 in 1980. This makes Singapore's proportion of workers in information occupations comparable to such countries as West Germany, Sweden, and the United Kingdom (Kuo and Chen, 1985, pp. 9 and 29). Therefore, with a plurality of its occupations in the information subsectors, Singapore for our purpose is assumed to be an information economy and society.

Lack of Information-Age Urban Planning

While there have been a few exceptions (Savitch, 1987, Perloff, 1980), on the whole there seems to be relatively little in the literature and practice of metropolitan planning that provides knowledge and guidance for planning cities in today's age of information, services and post-industrialism. The need for a more operational understanding of information-era metropolitan planning has been the motivating force for this preliminary analysis of Singapore's policies and planning.

Several years ago, when attempting to conceptualise a metropolitan planning strategy for the Seoul, South Korea, urban region, it first became clear to me that documentation about information-age metropolitan planning was at best quite sparce (Corey, 1982). The writings of Jean Gottmann about these issues have been useful to my thinking; he has formulated his observations about information and knowledge-driven urbanisation under the concept of the "transactional city" (Gottmann, 1983). This notion makes an important contribution to the substantive understanding of the information-age metropolis and its functions. What is needed still is a better understanding, indeed a working understanding, of how to *plan* an information-age metropolis (Corey, 1987). This essay is a preliminary effort toward meeting the need for more insight into such planning. By learning something of Singapore's planning experience and results, it is intended that analogous, but tailored information-age urban planning action might be stimulated elsewhere.

PROGRAM PLANNING MODEL

The conceptual framework of the Program Planning Model (PPM) was used in the design of this overall research project. It is also adopted here as the framework for reporting the preliminary observations drawn from Singapore's planning experience (Delbecq and van de Ven, 1971; van de Ven and Koenig, 1976). PPM is a seven-element procedural and planning-process model. It is premised on behavioural science research findings that emphasise the importance both of using experimental designs in plan implementation and in differentiating the principal actors and their respective special contribution to each phase of planning. The model has been used before to analyse policies and plans (Corey, 1984). PPM was judged to be appropriate for the Singapore case because Singapore policy and program planners have approached their task from an experimental and developmental perspective.

PPM Stage 1: The Initial Mandate

The dominant charter to plan in Singapore comes, for the most part, directly from the government to its bureaucracy. The mandate for the planned transformation of Singapore into an information-age metropolis ultimately stems from the collective will of the people for survival and prosperity. This drive for national survival has been instrumental both in PAP's return to power in each of the country's seven general elections and in the government's pragmatic strategies in planning and implementing its programs. Thus the mandate for development planning in Singapore derives from the citizens through the electoral process. In turn, the PAP governments have executed their mandates over the years through a public bureaucracy composed of administrative departments and statutory boards (e.g. Housing and Development Board).

From the earliest years of the Republic, coordination of Singapore's development has been taken at the highest levels of government. Action agencies, in the form of the departments and boards, are delegated the responsibility of designing and/or implementing programs and projects each of which is intended to contribute to one of the country's development strategies. However, second only to the electoral bond between the citizenry and the government, the ultimate *operational* mandate for Singapore's development has come from the Prime Minister. Since 1959 the operational mandate giver has been Prime Minister Lee Kuan Yew.

Characteristically, the method used at this stage in planning Singapore's development program is for the cabinet-level minister of a department to issue a strategy statement in a public forum (e.g. an annual Budget Statement). In turn, the media and government organs (e.g. *Mirror*) disseminate more detailed information on that planned development-program initiative. The cabinet and department minister or statutory-board head then monitors progress on program objectives during implementation.

PPM Stage 2: Problem Exploration

How have problems and needs fitted into Singapore planning? Problems, as identified by the elected government leadership with support from the bureaucracy, fuel Singapore's development response. Problems that confronted the new state of Singapore included:

- Unemployment (13.5 per cent in 1960),
- High birth rate (e.g. 3.2 per cent in 1960),
- Housing shortages,
- Political instability,
- Labour unrest, and
- Lack of infrastructure.

In general, nation building was the order of the day. Its ethnic and racial mix of numerically-dominant Chinese along with Malays, Indians and Europeans did not make the task a simple one. One of the key issues in the early 1960s was the conclusion that entrepot trade would not provide jobs in numbers sufficient to employ the growing population coming into the workforce. This stimulated policies and plans for the industrialisation of Singapore's economy (Tan and Hock, 1982).

Singapore's government generally has taken a strong rational-technical approach to planning. Its *People's Plan* states "Our aspirations must be built on hard facts: we can build on nothing less firm" (State of Singapore, 1961, p. 12). This bias toward planning from facts has produced useful data sources and published surveys (e.g. Ministry of Trade and Industry, 1985). One of *the* early and continuing measures for assessing the reduction of Singapore's problems has been annual per capita gross national product. This measure has been seen as the means of assessing change in real income, standard of living and quality of life.

Today, the principal problems being addressed by Singapore's policy and program planners have to do with *restructuring the economy* into even higher-technology manufacturing and higher-level services. Thus in the late 1980s the problems revolve around competitiveness, productivity, and the shifting of appropriate enterprises from public-sector management to private. With 84 per cent of the populace living in government-built flats and 72 per cent of these owning their flats in the Housing and Development Board new towns and projects (Information Division, 1986, p. 135), even housing problems have shifted to more concern for amenities, size and appointment of facilities. Many of Singapore's development problems have become less quantitative and more qualitative. For example, rather than the earlier concern for birth control, the contemporary concern has shifted to getting graduate women to marry and have children with men who are better educated. The projected demographic shifts to an older population relative to the working-age population represents another new and growing problem.

Government leaders and policy planners early conceived of Singapore's development within the context of the latest knowledge on modernity and contemporary urbanisation. Then Foreign Affairs Minister S. Rajaratnam envisioned Singapore's future as *a global city*; that "is the city that electronic communications, supersonic planes, giant tankers and modern economic and industrial organisation have made inevitable" (Rajaratnam, 1972, p. 3). Goh Keng Swee (1973, p. 51), often attributed as Singapore's early chief economic strategist, opined that in Asia the city has a modernising role to play. He admitted that this role was not well understood, but one that would be increasingly important in the future. The rational and experimental approach of Singapore's policy planning over the last generation represents a rich source of lessons for others as they grapple with harnessing post-industrial and information-age forces. Singapore is atypical in many respects, however it merits attention by urbanists and urban planners because it is one of the earliest metropolitan-scale places to use comprehensively-planned strategies and interventions in an attempt to move its economy and society deliberately into the information age. Also the scale of Singapore is relatively more manageable. And because its data are bounded and collected to its national borders, most statistical analyses enable generalisations for its entire urban system.

PPM Stage 3: Knowledge Exploration

Experts with state-of-the-art knowledge are the main actors on this stage; especially external experts and consultants. Since its inception as an independent entity Singapore policy makers and program planners constantly have scanned the external environment for innovative and effective solutions to its problems and for its development. Also, civil servants and students are sent abroad to return with needed training and education. This transfer of ideas and expertise into Singapore has played a significant role in the planning of its development programs. To this day, such technology transfer continues to be important. A few examples of focusing knowledge on selected Singapore issues and problems follow.

Technical assistance from the United Nations in the 1960s played important roles in, for example, Singapore's industrialisation drive (United Nations Industrial Survey Mission, 1963) and in its physical structure planning strategy (Doubai, 1971). Urban planning knowledge from overseas also reinforced Singapore's need for *flexibility* in planning. Given Singapore's extreme vulnerability to external influences and pressures, the city's leaders and planners must have a planning framework that facilitates changes in policies, programs and priorities. Planning technology transfer from abroad gave Singapore its long-range concept plan (Olszewski, 1972) and its flexible style of "action-planning" (Koenigsberger, 1964). External expertise was also used to assist in analysing the feasibility of building a mass rapid transit system (Hansen, 1980, Wildermuth, 1980).

PPM Stage 4: Proposal Development

The outcome of this stage is a strategic-level concept that requires further refinement and microdesign before its intent can be translated into operational programs. The elected elite and the bureaucratic elite generally are the key actors in this stage.

A series of proposals and concept plans is introduced below in chronological order. This sequence was constructed to trace some of the major milestones in Singapore's move to international industrialisation and later to upgrade its economy from promoting "brain services" to fomenting a "second industrial revolution" aimed at realising an information economy based on "information technology."

The People's Plan, 1961. To address its severe problems of unemployment, high rate of population growth, political instability and the need to create new jobs, Singapore's post-colonial government (self-government began 3 June 1959) developed and implemented a program of rapid industrialisation. From this earliest Singapore government, the *creation of jobs* has been a paramount task. In a Ministry of Culture publication series entitled "Towards Socialism," the 1961 *The People's Plan* states that "More jobs and more goods are the aims of this People's Plan" (State of Singapore, 1961, p. 9). Employment and the resultant earnings, especially as savings and investment, were seen as key means to achieve a better quality of life and standard of living for Singaporeans. These fundamental aims have been the driving forces for Singapore's development and planning from the early days of an autonomous Singapore up to the present time.

In the 1961 *People's Plan*, the government stated its role in mobilising Singapore's development as follows; this was part of Singapore's *first industrial revolution*:

> . . . Our sense of social justice which makes us demand social services must spur us to greater efforts in planning and creating our Industrial Revolution. Government can stimulate and guide this Industrial Revolution . . . (State of Singapore, 1961 p. 9)

The Singapore government's role is that of teacher and leader. In partnership with the private sector and the citizens of Singapore, the government sees itself as *the* actor whose role it is to provide the infrastructure and climate that will attract investment - both financial and technological - that is required to develop Singapore. *The People's Plan* of 1961 states "Economic development is now the responsibility of us all and not just a few people at the top" (State of Singapore, 1961, p. 1). To this day, the main actors in Singapore's development continue to work toward full realisation of this statement. The government and its bureaucracy have been, and remain *the* lead actors in Singapore's planning.

Development Plan, 1961-1964. Also in 1961 the State of Singapore published its *Development Plan 1961-1964.* Its main objective was to increase "employment opportunities for those who would be entering the labour market each year" (p. 33). Singapore's rapidly-growing population - then about 4 per cent annually - needed many new jobs, thus economic development took the lion's share of this plan's expenditure. Much of this expenditure was planned to generate revenue and thereby be *self-supporting*. These projects included power, water, gas, housing and port development. Finally, this plan explicitly manifested an enduring attribute of planning in

Singapore; that is its *flexibility*. It was stated that "The Plan is not rigid and is subject to periodic re-appraisal" (State of Singapore, 1961, p. 39).

A Proposed Industrialisation Program, 1963. In 1963 a mission of the United Nations (UN) produced a paper for the use of the government of Singapore. It was entitled, "A Proposed Industrialisation Program for the State of Singapore." Singapore's government commissioned this study to hasten the *expansion of manufacturing industries*. The UN analysis determined that 214,000 new jobs would be needed between 1961 and 1970. At the time of that study manufacturing industries were a rather "insignificant" part of Singapore's economy; 61,000 persons were employed in manufacturing industries from a total workforce of 471,000; "the output of the real manufacturing industries accounts for only 14 per cent of the total gross domestic product" (United Nations Industrial Survey Mission, 1963, p. ii).

Indeed, the UN study concluded that the reasons for a declining trend in Singapore's manufacturing included among others: *questions of political stability and uncertainty*, and *unfavourable industrial relations* (p. iii). To this day, monitoring and responding to these concerns remain key to the continuing success of Singapore's development.

The UN industrialisation program included *recommendations* for: an intensive export drive; technical education and training to strengthen the workforce; wages and productivity, i.e. reduce total unit cost by encouraging two or three shifts; attract foreign manufacturers, managers and capital; keeping a steady flow of local entrepreneurs and local capital from entrepot trade and commerce into manufacturing industries; insuring an active role for government in the promotion and nurturing of industrialisation; and limiting the role for government in direct participation in industrialisation. In addition, the proposed program recommended many financial incentives and tax measures designed to make investment in Singapore attractive. Because Singapore's manufacturers had been plagued by too many strikes, a dramatic improvement in *industrial relations* was also recommended. Employer and union cooperation was seen as essential to success in Singapore's future economic development. Tying wage increases to higher labour productivity, encouraging labour incentive systems, linking efficiency of manufacturers to larger profits, and government support for the modernisation of small industries were all recommended.

The UN study went on to discuss ways that Singapore might organise itself (e.g. the Economic Development Board) to enhance and promote its industrialisation policy. Selection criteria were set out to guide the development of specific industries in Singapore. Recommendations were made for: ship building and ship repairing; metal and engineering industries, electrical equipment and appliances; and chemical industries. Also introduced was a crash program designed to address the then immediate problem of unemployment. In hindsight, it appears that the UN proposal for an industrialisation program built the policy and planning foundation that has been drawn upon by Singapore decision makers for over twenty years of intensive economic expansion, physical development, and success in those initiatives.

Toward Higher-Level Practical Skills, 1970. In his 1970 budget speech, Finance Minister Goh Keng Swee noted that in the mid-1970s and beyond, Singapore expected to have a major problem of being able to supply investors with skilled personnel with *scientific and technological knowledge* and experience. He proposed addressing this problem initially by pursuing the practical approach of exploiting Singapore's advantage of "being able to produce or import engineers, technicians, craftsmen and skilled workers capable of performing the most exacting standards of work but at a much lower cost than the advanced countries could" (Goh, 1972, p. 276). The involvement of Singapore's universities and other educational and training institutions in the city-state's development thus became incorporated as a high priority for Singapore in the 1970s.

Development Plan for the 1970s, 1972. Under conditions of nearly full employment (unemployment then was 4.8 per cent) by 1972, Singapore's development policies and planning turned to a more sophisticated *high-technology and "brain services" approach.* This is an extension of earlier industrialisation policies. The need for a "Development Plan for the 1970s" was introduced in the 1972 budget speech by the Minister for Finance Hon Sui Sen (1972). It typifies Singapore's approach to development-policy planning for the promotion of modernisation. The plan concept was based on five elements: (1) manpower, (2) science and technology, (3) promotion of communications, transport and tourist infrastructure, (4) taxation incentives, and (5) brain services. The Plan's:

> . . . goal is to transform Singapore within ten years, at an economic development growth rate of 15 per cent per annum, into a regional centre for brain services and brain service industries. Success will depend upon the quality of our people and their capacity to improve and innovate through education, training and experience. This is an expensive process which can be carried out only if

> population growth is kept down to almost Z.P.G. (Zero Population Growth). Then we can better afford to attract expertise and technology from the outside world. (Hon, 1972, p. 23)

As to several of the plan elements, the Economic Development Board was geared to attract high technology industries and *brain services*. The latter included: "engineering and consultancy, marketing, physical planning and development, international sales and servicing, education, medicine and especially, banking and financing services" (Hon, 1972, p. 23). The major objective of promoting these brain services ultimately was to develop Singapore into an international banking and financial centre.

The *infrastructure element* (i.e. port facilities, airports and related services) of the development plan, among other outcomes, included studying the feasibility of a mass rapid transit system, and establishment of a statutory corporation to operate Singapore's *telecommunication services*.

As always in Singapore planning, *flexibility* was central to this plan for the 1970s. "The development plan will be supported by a program for public sector action, which will not be tied down to a rigid framework but will be flexible enough to accommodate contingencies and unforeseen circumstances" (Hon, 1972, p. 20).

Corrective Wage Policy, 1979-1981. During 1979, the government took "a bold economic step" (Goh, 1979, p. 48). As part of its *restructuring of the economy*, it extended its policy of promoting higher wages. Wages were increased dramatically over a three-year period. The intent of the corrective wage policy was to reduce inefficient labour users, to reduce the use of guest workers from abroad, to shift workers into higher-priority industries that require higher-skilled personnel and produce higher-value outcomes, and "to induce replacement of labour with productivity-raising machinery" (Rodan, 1985, p. 17). This intervention has been labeled the *"second industrial revolution."* The move to phase out low-skill, labour-intensive industries was necessary because of protectionism by advanced economies of low-value trade and because of competition from developing economies.

The *National Wages Council*, composed of representation from labour, employers and government, was the mechanism used to make higher-wage recommendations and to set wage guidelines. Also, "crucial to the success of this program of restructuring our economy is not only the training of large numbers of technicians but also upgrading the level of their expertise" (Chua, 1979, p. 16). Manpower development, and the role of training institutes and tertiary education thereby assumed an even more significant and active role than in earlier phases of Singapore's development. This wage policy, along with increased mechanisation, automation and computerisation, contributed to the overall goal of improving productivity.

Economic Development Plan for the 1980s. Introduced in outline form in 1980 (Goh, 1980, p. 3) and elaborated in 1981 (Goh, 1981, Appendix I), Singapore's government prepared a ten-year "Economic Development Plan for the Eighties." Its principal objective was "to develop Singapore into a modern industrial economy based on science, technology, skills and knowledge" (Goh, 1981, p. 5.).

This plan was similar to the one outlined in 1972 (Hon, 1972), in that it had five elements or "pillars of growth:" (1) manufacturing; (2) trade; (3) tourism; (4) transport and communications; and (5) brain services (which included: computer, financial, medical and consultancy services). The strategies that drove the plan were: a continuation of the corrective wage policy and wage increases announced in 1979, an increase in skills training and practical-education upgrading, and tax incentives to achieve automation, computerisation, and research and development.

Toward New Economic Directions, 1985 and 1986. In the Budget Statement for 1985, Finance Minister Tony Tan reviewed Singapore's economy for the period 1980 through 1984. He concluded that "the last five years have been years of exceptional prosperity for Singapore and Singaporeans. Singapore's gross national product (GNP) grew at an average rate of 8.5 per cent per annum in real terms in the period 1980-1984" (Tan, T. 1985, pp. 1-2). He also stated that economic growth for 1985 "is likely to be lower than the 8.2 per cent recorded in 1984. But growth there will be . . ." (p. 1).

One year later, after Singapore's economy had declined by 1.8 per cent in 1985, new Finance Minister Richard Hu stated that "restoring economic growth must be the priority for the coming financial year" (Hu, 1986, p. 1). Restoring economic growth thus became *the* short-range objective for 1986. Revised longer-range objectives and initiatives were detailed in *The Singapore Economy: New Directions* (Economic Committee, 1986) and in a spurt of new sector plans that were prepared in response to this economic slow-down. This was the first decrease in economic growth for Singapore in twenty years, and it launched the government into a flurry of plan making. Some of these included: a national plan for information technology, a tourism plan, a plan for conservation in the city's Central Area, and among others, a property-sector plan (Property Market Consultative Committee, 1986).

The keystone of Singapore's 1985-1986 round of planning however, was the "New Directions" report prepared by the Economic Committee (1986). The committee was chaired by Acting Minister of Trade and Industry, Lee Hsien Loong, the son of Prime Minister Lee Kwan Yew. The report identified the causes of Singapore's recession, including global conditions, reduction in United States' economic growth, and economic problems in the southeast Asian region. Causes of the recession internal to Singapore were seen to be:

- A loss of competitiveness; this includes total wage costs that rose twice as fast as productivity.
- Too heavy a reliance on the construction industry; "we should have become alarmed that a third of our economic growth each year derived from construction" (p. 51).
- The country was over-saving and over-investing in construction.
- There was seen to be rigidity in the domestic economy in that there was no built-in mechanism to reduce total wage costs when the economy and productivity contract.

The "New Directions" report also proposed future policies for economic growth and strategies by sector. The *policy areas* were: taxation, wages, manpower supply, education and training, productivity, entrepreneurship, local businesses, promotion of services, high technology and research and development, and information technology. The economic *sectors* addressed by the recommended future strategies included: manufacturing, banking and financial services, services, tourism, construction, commerce, and international trade.

Through the years, the ultimate goal of Singapore's development planning has remained constant; *job creation* for the realisation of a high quality of life and a commensurate living standard. One of the *original goals* of the People's Action Party (PAP) was:

> . . . to establish an economic order which will give to all citizens the right to work and full economic returns for their labour and skill; to ensure a decent living, and social security to all those who through sickness, infirmity or old age can no longer work. (Tan, 1982b, p. 22)

Even after the economy's most stressful time in Singapore's modern period, the quality of life and standard of living goal was reaffirmed:

> . . . The end is to provide Singaporeans with the opportunity to earn the standard of living which they deserve, by being well organised, by working in productive, well-paid jobs. To do so, we must make Singapore a place where businesses thrive, opportunities abound, and honest fortunes can be made. (Lee, 1986, p. 56)

Each of these concepts, policies, plans, programs and studies contributed principles and practices that Singapore policy makers and planners have used and perfected from many years past to the present. The aggregate of these and other practices represents much of Singapore's planning for its external and internal problems. The primary actors in this stage are mainly elected government ministers.

PPM Stage 5: Program Design

The chief actors at this stage generally come from within the public bureaucracy. They are professional program planners and technicians who translate proposed concepts from the previous stage into operational programs and activities. While there are many Singapore development programs to be described and analysed for this research (e.g. the activities of: the Urban Redevelopment Authority, the Singapore Tourist Promotion Board, the Economic Development Board, the Jurong Town Corporation, the Planning Department, the Development and Building Control Division, and the Mass Rapid Transit Corporation, among others), space limitations here permit the treatment of just one emergent program. Accordingly, the forces that have produced the recent information technology policy are explored.

Information Technology (*IT*). Singapore policy makers and planners see the IT policy both as a means for *improving productivity* and as a *growth industry* itself. In addition to public-sector initiatives, the private sector is expected increasingly to contribute to the IT strategy. Further, "companies in both manufacturing and services need to exploit fully the advances in IT" (Economic Committee, 1986, pp. 152-53). The promotion of services, including information services (Langdale, 1984), forms part of Singapore's latest economic policy:

> . . . We have a comparative advantage in exporting services greater than our advantage in exporting goods. We have comprehensive communications and telecommunication links to the rest of the world . . . . Without underrating the

> importance of manufacturing to the economy, we can safely predict that services will be a leading growth sector in Singapore in years to come. (Lee, 1986, p. 63)

As a result of past initiatives into higher-technology manufacturing and higher-level services, by the early 1970s Singapore was well on its way to becoming an information economy. But what is this type of economy?

The *information economy* consists of functions and activities that are:

> . . . supported by the establishment of industries and markets wherein the necessary technical infrastructure are produced and information as a commodity is sold respectively. As these activities expand in scope and volume, there is a concomitant increase in the share of GNP arising from the value added which originates from the production and distribution of information goods and services. (Jussawalla and Cheah, 1983, p. 162)

Using Singapore's input-output table for 1973, Jussawalla and Cheah (1983, p. 168) concluded that indeed "the results of our empirical study of Singapore's information sector . . . suggest that Singapore is proceeding towards an information economy." Their study began the important process of accounting for the product generated by the information sector of Singapore's economy. The need for the *measurement and empirical analysis* of service economies in general, and information economies in particular, has been noted widely (Gottmann, 1983, Shanahan, 1985, Shelp and Hart, 1986). Jussawalla and Cheah (1983) found that the information sector accounted for 24 per cent of Singapore's gross domestic product in 1973. Also, they concluded that:

> the public sector appeared to be the most information-intensive with approximately two-thirds of all government services being information-based. Public bureaucracies in administration, planning and development agencies accounted for 49 per cent of total value added. (Jussawalla and Cheah, 1983, p. 168)

The importance of the public bureaucracy, in turn, has been used to spearhead Singapore's computerisation drive. Following directly from its "development strategies for the eighties" to automate and computerise (Goh, 1980), Singapore's Civil Service has become a role model for other sectors of the economy to emulate and thereby become more productive (Wong, 1986). The National Computer Board (NCB) has been assigned the principal role of moving Singapore into the information age (National Computer Board, 1984-1985). The formation of the NCB was preceded by Singapore's government, appointing, in March 1980, a high-level ministerial Committee on National Computerisation. Six months later, this committee recommended the formation of the NCB to:

- promote the training of more computer professionals for the industry,
- implement computerisation of the Civil Service and other Public Sector organisations,
- promote software development for export.
  (National Computer Board, 1983, p. 1).

By November 1985, the NCB and other government agencies and institutions produced a conceptual *national plan for information technology*. In Singapore, information technology (IT) included "the use of computer, telecommunication and office systems technologies for the collection, processing, storing, packaging and dissemination of information" (National IT Plan Working Committee, 1985, p. i).

As information technology strategy is considered so important both to Singapore's immediate and long-term economic development, that it was included as one of the principal policies for growth in the recent landmark "New Directions" report (Economic Committee, 1986). Again, this effort was commissioned in March 1985 (Tan, 1985) to conduct a mid-term assessment of Singapore's economic development plan for the 1980s (Goh, 1981, Appendix I). The review was conducted throughout 1985, a year that saw Singapore's first economic decline in twenty years. In real terms, gross domestic product decreased by 1.8 per cent (Ministry of Trade and Industry, 1985, p. 3). In the midst of the recession, the government saw the IT strategy as "a silver lining in the economic dark cloud." Senior Minister S. Rajaratnam observed:

> . . . an information economy . . . is an ideal economy for Singapore. With minor exceptions, information operations take up very little space. They consume very little energy relative to the profits they yield and offer scope for the setting up of small enterprises . . . Most important of all, its operations are dependent on brain resources (which are inexhaustible) rather than on natural resources. Also

> information products are less subject to tariff barriers than industrial products in a competitive world. (Rajaratnam, 1986, p. 17)

At the IT exhibition "Singapore Informatics 1986" in December of that year, the outline of the national IT plan received a great deal of publicity and dissemination to the public (The Straits Times, 1986, p. 14). The plan includes seven strategic building blocks: (1) IT manpower; (2) IT culture; (3) information communication infrastructure; (4) applications; (5) climate for creativity and entrepreneurship; (6) IT industry; (7) coordination and collaboration (National Computer Board, 1986). Key actor-agencies and their roles in Singapore's IT development include:

- *National Computer Board* has the role of promoting computers and development of the computer-services industry.
- *Telecoms'* role is to develop the telecommunications infrastructure and to work with the private sector to exploit IT (Chia, 1986).
- *Economic Development Board*'s role is to promote investment in computer industries and services.
- *National University of Singapore* is responsible, in part, for the development of IT manpower.

The burden of the plan is to "provide all organisations in Singapore a clear view of the future" (National IT Plan Working Committee, 1985, p. 46) and to realise synergy by reducing fragmentation among existing agencies with responsibilities for implementing parts of the IT strategy.

Housing and Development Board. Additionally, the housing, physical planning and urban spatial structure programs in Singapore's development have been well documented. Housing is one of Singapore's most famous planned "program design" and program implementation examples. The highly successful housing and new town development program was thoroughly researched and assessed (Wong and Yeh, 1985).

PPM Stage 6: Program Implementation and Program Transfer, and Program Evaluation: Spans PPM Stages 4 through 6

As part of the budget-statement process, Singapore's economy is monitored annually. These yearly (and quarterly) published surveys are widely publicised (e.g. Ministry of Trade Industry, 1985), and they serve as a kind of public evaluation of progress (or not) on broad national goals and sectors of the economy.

The program design and action agencies themselves generally are responsible for testing, implementing and evaluating their own development programs and projects. As noted above such agencies include: the Urban Redevelopment Authority; the Jurong Town Corporation, the Housing Development Board; Public Utilities Board; Telecoms and others.

However, the Planning Department, especially through its regularly updated "concept plan," identifies future development needs and directions. Ultimately such issues may be programmed into action, but by one or several development agencies.

CONCLUSION

> . . . Singapore is quite unique in the sense that it is moving to become an information society, judging from the proportion of information workers in the labour force, while still at the prime of industrialisation. (Kuo and Chen, 1985, p. 14)

Because of the many ideosyncratic characteristics of Singapore, it is clear that there cannot be a direct transfer of information-age urban planning technology from Singapore to other metropolitan areas. However, by extracting *lessons* from Singapore's long-term experience in planning initiatives and programs to promote high-technology manufacturing and higher-level services, one can draw inferences about how these lessons might begin to inform information-age metropolitan planning in, for example, North America.

Singapore's government leaders, in reflecting on the reasons for Singapore's development successes have provided us with some lessons:

- Singapore's success has been due primarily to the trust that has been built up over a twenty-six year period of nurturing. Such lack of trust is a root cause of Europe's unemployment problems where a constant change of government every four or five years has undermined public trust (Goh, 1986).
- The key to Singapore's long-term success in its economic restructuring strategy is not higher wages, nor fiscal incentives. It is manpower development (Goh, 1981, p. 1).
- Singapore will still be able to attract some foreign investments, because of the very strong assets built up over the years, namely political and economic stability, sound infrastructure, and a capacity for hard-headed and practical adjustments to changing circumstances (Goh, 1981, p. 11).

- All three races (Chinese, Malay and Indian) of people respect authority. The government is seen as having integrity, being clean and noncorruptible (Low, 1986).
- Singapore is realistic in its expectations and assessment of the difficulties ahead (Goh, 1981, p. 24).
- That Singapore has done well in the face of an inhospitable economic environment is due to sheer hard work, grit, determination and adaptability on the part of its workers and managers, and to the good rapport built up over many years between our unions, employers and the government (Tan, 1982, p. 17).

In addition to being *flexible* in planning policies and programs, there is another lesson for us from Singapore. Planners of other metropolitan areas should seek and develop *niches*, especially economic, job-creation niches. By being sensitive and systematic in scanning the external and internal environments of our own metropolis, we may be better positioned to encourage some activities and to discourage others. The very act of seeking particular kinds of future outcomes based on knowledge of trends and innovations in information-age economies and societies, combined with empirical and marketing intelligence gathered on the potentials and needs of our cities and their people, should be an aid to developing a new and more appropriate approach and practice of metropolitan planning.

If we are to be competitive and successful in this changed world, we need to probe and plan the future unfettered from some of our comfortable and past analyses and practices. For example,

> Once we accept the fact that today's economy is not only very different from the kind of economy we are accustomed to thinking about, but that the old industrial economy will never return, the task of developing appropriate public policies will be much easier. (Shelp and Hart, 1986)

The information age no longer is in the future. It is not coming, it is here. Is it not now time that our metropolitan planning better reflects today's changed and changing realities?

Acknowledgements

This research was supported by the Fulbright Research Program. It enabled me to be a 1986 visiting scholar at the Institute of Southeast

Asian Studies (ISEAS) in Singapore. Appreciation for special support of the research is expressed to K.S. Sandhu, Sharon Siddique, Goh Hup Chor, Johan Ole Dale, Cheok Yen Aik, Loh Swee Seng, John Keung, Lau Woh Cheong, Loh Chee Meng, Tan Ban Seng, Chia Choon Wei, Lily Lu, Ong Choon Hwa, Richard Gong, Arthur Vaughn, Mark Chan, Jeanine Kleimo, Robert Fletcher, and the staff of ISEAS.

REFERENCES

Chia, C.W. (1986) "Economic aspects of the information revolution: the Singapore experience," in Jussawalla, M. *et al.*, (eds.), *The Passing of Remoteness? Information Revolution in the Asia-Pacific*, Singapore: Institute of Southeast Asian Studies, 42-50.

Chua, S.C. (August 1979) "Second industrial revolution," *Speeches*, 3(2): 15-17.

Corey, K.E. (1982) "Transactional forces and the metropolis," *Ekistics*, 44: 416-23.

Corey, K.E. (1984) "Qualitative planning methodology: an application in development planning research to South Korea and Sri Lanka," *Development Planning Review*, 3(2 and 3): 1-14.

Corey, K.E. (1987) "The status of the transactional metropolitan paradigm," in Gapper, G. (ed.), *The Future of Winter Cities*, Newbury Park, California: Urban Affairs Annual, Sage Publications.

Delbecq, A.L. and van de Ven, A.H. (1971) "A group process model for problem identification and program planning," *Journal of Applied Behavioural Science*, 4: 466-92.

Doubai, A. (1971) "Singapore's United Nations-assisted state and city planning project," *Singapore Institute of Planners Journal*, 1: 4-8.

Economic Committee (1986) *The Singapore Economy: New Directions*, Singapore: Ministry of Trade and Industry, Republic of Singapore.

Goh, C.T. (July 1979) "Restructuring the economy through higher wages," *Speeches*, 3(1): 46-48.

Goh, C.T. (5 March 1980) "We must dare to achieve," *Budget Speech 1980*, Singapore: Information Division, Ministry of Culture.

Goh, C.T. (6 March 1981) "Towards higher achievement," *Budget Speech 1981*, Singapore: Information Division, Ministry of Culture.

Goh, C.T. (2 November 1986) From a television broadcast on Singapore Broadcasting Corporation, "Insuring your future," a speech the General Insurance Association of Singapore.

Goh, K.S. (1972) *The Economics of Modernisation and Other Essays*, Singapore: Asia Pacific Press.

Goh, K.S. (August 1973) "Cities as modernisers," *Insight*, 46-50.

Gottmann, J. (1983) *The Coming of the Transactional City*, College Park, Maryland: University of Maryland, Institute for Urban Studies.

Hansen, K.R. (September 1980) "Singapore's transport and urban development options," Final report of the MRT Review Team, Singapore: Republic of Singapore.

Hon, S.S. (7 March 1972) *Singapore: Economic Pattern in the Seventies*, Singapore: Ministry of Culture.

Hu, R.T.T. (7 March 1986) *Budget Statement 1986*, Singapore: Information Division, Ministry of Communications and Information.

Information Division (1986) *Singapore Facts and Pictures 1986*, Singapore: Ministry of Communications and Information.

Jussawalla, M. and Chee-Wah, C. (1983) "Towards an information economy: the case of Singapore," *Information Economics and Policy*, 1: 161-76.

Koenigsberger, O. (1964) "Action planning," *Architectural Association Journal*, 79(882): 306-12.

Kuo, E.C.Y. and Chen, H.T. (1985) *Towards an Information Society, Changing Occupational Structure in Singapore*, Singapore: University of Singapore, Select Books for the Department of Sociology, Sociology Working Paper.

Langdale, J.V. (1984) *Information Services in Australia and Singapore*, Kuala Lumpur and Canberra: ASEAN-Australia Joint Research Project,, ASEAN-Australia Economic Papers No. 16.

Lee, H.L. (Jan.-Feb. 1986) "Singapore's economic policy: vision for the 1990s," *Speeches*, 10(1): 56-82.

Low, S.S. (11 November 1986) Personal interview with the Chief Planner, Planning Department, Singapore Ministry of National Development.

Ministry of Trade and Industry (1985) *Economy Survey of Singapore 1985*, Singapore: Republic of Singapore.

National Computer Board, (1983) "Government takes bold stand," *IT Focus*, Singapore: National Computer Board.

National Computer Board, (1984-1985) *Moving Singapore into the Information Age*, Year Book FY 84/85, Singapore: National Computer Board.

National Computer Board (1986) "National IT Plan," Singapore: Republic of Singapore, 1-8.

National IT Plan Working Committee (30 November 1985) *National IT Plan, A Strategic Framework*, Singapore: National Computer Board.

Olszewski, K.F. (1972) "The principle of flexibility as applied to the structure of the Singapore concept plan," *Singapore Institute of Planners Journal*, 2: 5-17.

Organisation for Economic Cooperation and Development (OECD) (1981) *Information Activities, Electronics and Telecommunications Technologies*, Vol's. I and II, Paris: OECD.

Perloff, H.S. (1980) *Planning the Post-Industrial City*, Washington, D.C.: American Planning Association.

The *People's Plan* (12 April 1961) Towards Socialism, Vol. 4, A Ministry of Culture Series, Singapore: State of Singapore.

Property Market Consultative Committee (February 1986) *Action Plan for the Property Sector*, Singapore: Ministry of Finance, Republic of Singapore.

Rajaratnam, S. (6 February 1972) *Singapore: Global City*, Singapore: Ministry of Culture.

Rajaratnam, S. (March-April 1986) "Information economy is ideal for Singapore," *Speeches*, 10(2): 17-25.

Rodan, G. (1985) *Singapore's "Second Industrial Revolution": State Intervention and Foreign Investment*, Kuala Lumpur and Australia: ASEAN-Australia Joint Research Project, ASEAN-Australia Economic Papers No. 18.

Savitch, H.V. (1987) "Post-industrial planning in New York, Paris, and London, *Journal of the American Planning Association*, 53(1): 80-91.

Shanahan, E. (27 October 1985) "Measuring the service economy," *The New York Times*, 4-F.

Shelp, R.K. and Hart, G.W. (23 December 1986) "Understanding a new economy," *The Wall Street Journal*, 20.

State of Singapore (1961) *State of Singapore Development Plan 1961-1964*, Singapore: Ministry of Finance.

State of Singapore (12 April 1961) *The People's Plan*, Towards Socialism, Vol. 4, A Ministry of Culture Series, Singapore: Ministry of Culture.

Straits Times, The (4 December 1986) "Seven-prong approach to National IT Plan," *The Straits Times*, 14.

Tan, A.H.H. and Hock, O.C. (1982) "Singapore," in *Development Strategies in Semi-industrial Economies*, Baltimore: The John Hopkins University Press, 280-309.

Tan, T.K.Y. (5 March 1982a) *Budget Statement 1982*, Singapore: Information Division, Ministry of Culture.

Tan, T.K.Y. (December 1982b) "PAP's objectives for future decades," *Petir*, No. 15: 14-22.

Tan, T.K.Y. (8 March 1985) *Budget Statement 1985*, Singapore: Information Division, Ministry of Communications and Information.

Teh, C.W. (1979) *Planews*, 6(2):12.

United Nations Industrial Survey Mission (1 September 1963) "A proposed industrialised program for the State of Singapore," New York: Commissioner for Technical Assistance, United Nations.

van de Ven, A.H. and Koenig, R. Jr. (1976) "A process model for program planning and evaluation," *Journal of Economics and Business*, 28(3): 161-70.

Wildermuth, B. (1980) "An assessment of the Hansen Review Team's Report on 'Singapore's Transport and Urban Development Options'," Singapore: Wilbur Smith and Associates.

Wong, A.K. (5 November 1986) "How changes have helped us," *The Straits Times*, 14.

Wong, A.K. and Yeh, S.H.K. (1985) *Housing a Nation: 25 Years of Public Hcusing in Singapore*, Singapore: Maruzen Asia for Housing and Development Board.

CHAPTER 4

SOUTH AFRICA: THE FAILURE OF FLAWED GOVERNMENT PLANNING

Francis. H. Horn
Emeritus Professor
The University of Rhode Island,
Albertus Magnus College
and
Pratt Institute
Providence, Rhode Island

FOREWORD

Unlike the many distinguished contributors to this volume honouring Peter Nash, I am not an architect, a geographer, or a planner. I am primarily a university administrator, having spent approximately fifty years in the higher echelons of college and university administration as dean, vice-president, and president of a number of institutions of higher education in the United States and abroad. Though not a scholar in the fields in which Peter and the other contributors are scholars, in my last position, at the leading private university in Taiwan, I was designated a "distinguished visiting scholar." I would argue, moreover, that I have been a planner, that any university administrator must be a planner, though not in the sense that schools of planning produce planners. But if planning involves establishing objectives, studying how best to meet them in the light of the specific circumstances, and then organising resources of people, materials, and finances to meet them, then a university dean or president is a planner.

In any case, Peter Nash served as a dean at The University of Rhode Island and at the University of Waterloo. Thus, he is a planner in the sense I've suggested, as well as in the more traditional meaning of the term. Since I appointed Peter to his first major administrative post, an account of how this happened may be appropriate to this volume and indicate to some extent my admiration and respect for him.

I had come to The University of Rhode Island as president in 1958, after having served as president of Pratt Institute, in Brooklyn, New York. Pratt had an outstanding School of Architecture. A creative faculty had begun to broaden the school's interest in architectural planning to include comprehensive town and city planning, and we established a master's degree program in the field.

Because my own interest in such planning had been aroused and stimulated by the Pratt experience, shortly after settling in at Rhode Island and searching for challenging and feasible new academic areas for a university of our size and resources, I proposed a new program in planning. The trustees approved, with the program to begin in 1963, and the search for someone to head the program began. Peter, then at the University of Cincinnati, had already distinguished himself as a planner and a geographer. His credentials easily made him the top candidate. In the interviews with Peter we were so impressed by his personal and professional qualifications that we not only offered him the chairmanship of the new Department of Community Planning and Regional Development, but also that of Dean of the Graduate School, a position then vacant. Peter accepted both positions and joined the University in the fall of 1963.

Our confidence in Peter was rapidly confirmed. He got the new program off to a flying start. He proved to be a person of sound academic ideas, and he possessed extraordinary creative ability, boundless enthusiasm, and almost unlimited energy.

The Planning Department was an exciting place to be. He recruited able faculty and attracted good students. Graduates of the program have been highly successful.

Peter was also successful in developing the Graduate School from its modest beginnings a few years earlier. Eventually, however, holding down the two demanding positions proved too much for even someone with Peter's ability and dedication, so he elected to concentrate on his professional interests in planning. But after seven years at URI, he left for new academic worlds to conquer north of the border. Though reluctant to see him go, his colleagues and the administration understood his reasons. The planning program at URI was on solid ground. Peter had built up widespread recognition for it. It was natural for him to seek out new challenges. These he found at the University of Waterloo. His subsequent career there as teacher, researcher, author, administrator and consultant is better known to his professional colleagues who are contributing to this volume than it is to me. His achievements are outlined in his listing in *Who's Who in*

America. It is an impressive record of the remarkable career of a true giant in a major academic and professional field.

I realise that what I have written thus far is not what the editors had in mind when they invited me to contribute to this volume of essays. I believed, however, that my best contribution to the purpose of the volume, to honour Peter, would be to pay direct tribute to him. This I have tried to do. But the editors wanted the volume to be a typical *festschrift*, with scholarly essays on some topic related to one of Peter's many interests.

Like Peter, I consider myself an internationalist, a "citizen of the world," so to speak. In fact, I am a registered "planetary citizen." I have lived in five foreign countries over a period of years, always tilling in the academic vineyard, and have travelled in more than seventy-five, in some of which I have done academic consulting. I drew plans for a small university in the former Himalayan Kingdom of Sikkim, now part of India. I headed an international commission which planned the King Abdul Aziz University in Saudi Arabia, to which I have returned several times for further consultation. I have also had experience with universities in the Far East, having served as chairman of trustees of the United Board for Christian Higher Education in Asia, with my final position at Tunghai University in Taiwan. The problems of any of these areas would make an interesting essay relevant to Peter's scholarly interests. But from the standpoint of concrete solutions to problems resulting from government planning, I believe the most complex situation, and therefore the most interesting, exists today in South Africa, from which I returned several months ago, although I have never lived there. In addition, the agony of South Africa is once again on the front pages of the newspapers, featured in our news magazines, and reported on our nightly television newscasts. Given the acceleration of violence in the country, the possible slide into civil war, and the involvement of the Western World in trying to avoid such a catastrophe, South Africa is likely to remain on the front pages, and its racial struggles continue to be of major interest to the English-speaking world and to the nations of Europe and Africa. Consequently, because of this intense concern with those struggles and their eventual outcome, because Peter's most abiding interest is in planning and its components in geography and demography, and because I regard the situation in South Africa as the most ambitious, though ill-advised, government planning program of the modern world, I have elected to deal with South Africa.

SOUTH AFRICA

First a few background facts. The Republic of South Africa, the official name of the country, covers an area of 426,000 square miles, approximately the size of West Germany, France, Italy, the Netherlands, and Belgium combined. The population in 1985 was a little less than 31,000,000, including the population of the four "independent" homelands. Without them, the population is approximately 24,000,000. Of the total, 4.5 million, 15 per cent, are white; 22 million, 73 per cent, black; 900,000, 3 per cent, Indian; and nearly 3 million, 9 per cent, "coloured." The coloured would be considered blacks in the United States or Canada. They are of mixed race descendants of the early white settlers, their slaves and the native Khoisan peoples known to Europeans as Bushman and Hottentots who inhabited the area when the white men first arrived. The current black population occupied the central and south-eastern parts of south Africa many hundreds of miles to the north and east of the early white settlements. Toward the end of the eighteenth century, conflicts between the blacks and the white farmers resulted in a long series of wars. Thus, the roots of the current struggle between blacks and white go back many years.

The Dutch were the first white settlers to arrive in what is now South Africa, establishing a permanent foothold in 1652. The English began coming toward the end of the eighteenth century, and after the territory was awarded to Great Britain by the Treaty of Vienna in 1814, they began arriving in large numbers. Today the white population of South Africa is divided between the Afrikaners, the descendants of the early settlers, and the English, with the language of each officially recognised. Although the Dutch were the first of the Afrikaner population, only about 40 per cent of that population is ethnic Dutch. Forty per cent are of German extraction, 15 per cent French, descended mainly from Huguenots fleeing Catholic persecution after the revocation of the Edict of Nantes in 1685, and 5 per cent Portuguese or of other European background.

The English gradually wrested control from the Afrikaners, known as "Boers" ("boer" is simply the Dutch word for "farmer"). The Boers, objecting to British policy of outlawing slavery and according legal rights to free blacks and coloured, and seeking to preserve their own lifestyle and freedom from English colonial domination, moved north out of the Cape Colony in the great Voortrekker exodus of 1835-1841, suffering serious losses on the way from attacking black tribes, including the massacre of women and children, a tragedy in their past history never forgotten by the Afrikaners. They went first into Natal, and then moved further north and west into what became the Orange Free State and the Transvaal, where the Boers established their own republics. Natal, also originally an independent Boer republic, was annexed by the British in 1843, the Orange Free State in 1871, and the Transvaal in 1877, but the latter's independence was restored in 1881.

In the nineteenth century, there was continuing conflict between the non-Afrikaner whites and the Afrikaners, especially after the discovery of diamonds in 1869, and of gold in 1886, brought a major influx of "outlanders," mainly British, into the predominantly Boer areas. The Boers reacted by declaring war on Great Britain in 1899. The resulting "Boer War" ended with a British victory, after bitter fighting, in 1902. In 1910, by consolidating the two British provinces of Cape Colony and Natal, and the former Boer republics of the Orange Free State and the Transvaal, the British established the Union of South Africa. These four major divisions remain the component states of South Africa today.

Although the Boers had lost the war, they have provided the leadership of South Africa ever since. The nation became independent in 1931, but remained part of the British Commonwealth. In 1961, however, it withdrew over the issue of apartheid and took the new name of the Republic of South Africa.

The conservative Nationalist Party, which came to power in 1948, following more liberal Afrikaner governments, established the concept of apartheid as the basic ideological and operational policy for the governance and development of the country. "Apartheid" simply means "separateness." The government determined that the whites were to be separated from the nonwhites both as to daily contact and as to residential living. This abstract basis of apartheid was further refined to divide the nonwhite population into the three racial groups of the blacks, the coloured, 90 per cent of whom lived in Cape Province, and the Indians, 80 per cent of whom lived in Natal.

By the Populations Registration Act, every resident of South Africa was legally and officially classified into one of these four racial groups, and issued identification documents. The policy further decreed that the different groups were to live in separate areas and to the maximum extent possible not mingle in the social, business, and political activities of normal everyday life. The foundations of this segregation by colour and race had already been laid before 1948. But it was not until after that date that apartheid in the present form was legalised.

The government planners turned their attention also to an even more complex aspect of apartheid, the separation of the black tribes.

Tribal enmities have existed for milleniums, and violence between the various tribes has been constant over the centuries. The great Zulu warrior, Chaka, in the early part of the nineteenth century, for example, rose to power by the slaughter of some two million other blacks before he himself was assassinated.

These tribal conflicts have continued to this day. Recognising these tribal threats to peace and order, and undoubtedly seeking greater control over the blacks, the government initiated legislation in 1959 to separate the tribes by establishing independent "bantustans," or tribal homelands, which were expected eventually to become national states. Homelands were planned for nine tribes, with the Xhosas to have two. In addition to the Xhosas, the tribes were the North Sothos, the South Sothos, Tswanas, Zulus, Swazis, Tsongas, Vendas, and Ndebeles. They were to occupy separate geographical areas, not necessarily all within one boundary. The largest, for example, Bophutatswana, has seven different noncontiguous areas. The homelands occupy some 13 per cent of the nation's area, and range in size from Quaqua's 500 square kilometres to Transkei's 45,000. The official language of each homeland is the tribal language. Government plans called for these tribal areas to be self-governing as to local concerns, and eventually all to be "independent" of Pretoria, the administrative capital. The legislative capital is Cape Town, and the judicial capital, Bloemfontein. To date only four independent homelands have been created: Transkei, in 1976, and Ciskei, in 1981, for the Xhosas; Bophutatswana, in 1977, for the Swanas; and Venda, in 1974, for the tribe of that name. The major tribe, the Zulus, with over six million members, are in KwaZulu, and it, together with Lebowa, have voted to remain part of South Africa. No final decision has yet been made about the four other tribal areas.

The so-called independent homelands consider themselves "countries" and designate themselves "republics." Although visas are required for foreign visitors, the "republics" are not recognised by governments other than that of South Africa, and thus have no diplomatic missions abroad, although they do maintain embassies and consultates in Pretoria and Cape Town, and have trade or other missions abroad. Bophutatswana has them in seven countries, including the United States.

Theoretically, all black South Africans were to be assigned to one of the tribal areas, have their principal residence and exercise their citizenship there - the right to vote on area matters but not on general South African issues. In practice, however, homeland residence proved an impossibility for the whole black population. Business, industry, and domestic service required the presence of too many nonwhites within a commuting distance of their place of work, although that distance sometimes became an unreasonable one in order to achieve the goal of separateness.

To meet this problem, the government established the concept of exclusively black townships. Blacks were not allowed to live in white - or Indian or coloured - areas where they worked, but were segregated in these townships, located at varying distances from the larger cities. Soweto, the largest, with a population of 1 1/2 to 2 million, is a suburb of Johannesburg. In a number of cases, the townships were created from scratch. In others, existing black communities were officially designated townships.

To enforce the separation of other nonwhites, the Indians and the coloured, from the white population, the government enacted another cornerstone of apartheid, the concept of "group areas." They are "designated and demarcated stretches of land (usually in urban areas) which can only be legally owned and occupied by people of a particular racial group." By the end of 1984, a total of 899 group areas had been proclaimed, 451 for whites, 326 for coloured, and 122 for Indians. There are no group areas for blacks as they are assigned to the homelands, although many live in the black townships. By the end of August 1984, a total of 126,176 families, over 800,000 persons, had been moved, only 2 per cent white, but 32 per cent Indian, and 66 per cent coloured. A total of 2,771 businesses had also been moved, over 90 per cent owned by Indians. Exceptions are made to the residence rules to permit nonwhites who are bonafide domestics of white families, to live in the homes where they work.

Undoubtedly one of the government's policies most criticised has been the forcible removal of families to achieve the goals of apartheid, both for group areas and for the homelands. By 1983, it was estimated that a total of 3 1/2 million persons had been moved, and another 1.8 million were under threat of removal. Over a million blacks have been moved as tenants on white farms, many of them, however, because their services were no longer needed. Almost 700,000 were moved when entire black townships were moved to the homelands. There is no doubt that the government has been relaxing the practice of forcible removal. In February 1985, it was stopped pending a study of the policy, and in May 1986, the government listed fifty-two black townships, previously slated for removal, where the residents were not to be removed.

With the reaction of the government to international condemnation and the imposition of sanctions by the United States, it is still too early to determine whether or not the government will reinstitute such harsher measures of apartheid. President Botha has made it clear that his government has no intention of ending the basic apartheid concept of residential separateness, however many other specific aspects of apartheid have been eliminated or were scheduled for elimination.

In this connection one must remember the long-time policy of the U.S. government for American Indians, forcing them to live on reservations and restricting their movements therefrom. Even today, the Bureau of Indian Affairs is in the process of forcibly moving against their will some 30,000 Indians in the southwest to keep the Hopis and the Navajos separated in their tribal areas, akin to the South African "homelands." Our own flawed record of racial justice in the United States should, without allowing us to condone the lack of it in South Africa, at least make us more understanding of the much more complex racial situation in that country, and less demanding of a quick end to racial injustic there.

Before considering some of the major issues in this complex racial situation, let me set forth my justification for commenting on South Africa in the first place, as I am not a specialist on the country. What follows is a layman's reaction based primarily upon a three-week visit in August 1986, supplemented by my reading about the country. This does not make me an expert on South Africa, but at least I have been there, and recently, which is important because of the changes which have occurred since 1984. As one of my professional colleagues on the trip stated: "I found that even informed Americans can't really understand what's going on in South Africa unless they've been there."

Most Americans have strong opinions about South Africa, but little actual knowledge concerning the situation there. They get their ideas from the limited and sensational news coverage on nightly television, or from pronouncements by radical South African black leaders like Archbishop Tutu, Mrs. Winnie Mandela, and Oliver Tambo, echoed by most black leaders in this country. Americans get almost no information about the moderate blacks, who constitute nearly three-fourths of the black population, or from their leaders like Zulu Chief Mangosuthu Butelezi, or the black mayors who travelled with us.

It is not only American college students who for the most part are ignorant of the real circumstances in South Africa. College trustees, who have voted divestment, city councilmen and state legislators, who have passed resolutions condemning South Africa, and members of Congress who have voted to override President Reagan's responsible veto of the sanctions bill are equally uninformed. Like most Americans, they are reacting emotionally, not rationally, to events in South Africa. Student construction of shanties on campuses across the land, for example, implies that most South African blacks live in shanties, which some do. But one of my colleagues on the trip who had been a Peace Corpsman in South America, in an interview after his return, stated that even the shanty town Crossroads "was not as bad as the 'Barrrios' I saw in Caracas." Yet students don't mount demonstrations for shanty dwellers in other parts of the world, or for the black Americans in urban ghettos or white Americans in rural Appalachia, many of whom live in circumstances as bad as the worst off blacks in South Africa. Students and the others who think of blacks in terms only of shanties, and conditions in shanty towns are bad, it is admitted, do not seem to realise that there is a growing black middle class who live in comfortable surroundings, and even a considerable number of black millionaires.

I would point out, incidentally, that in addition to benefiting from a good deal of reading about South Africa, my perspective was enhanced during the visit, and in sorting out my reactions afterwards, by my having lived in Egypt for three years, with many return visits there and to the other nations of North Africa, and by having spent over a month travelling in the black nations of Kenya, Uganda, Tanzania, and Ethiopia. My living abroad for nearly ten years, and my having travelled in most areas of the world outside of South America, also gave me a broad background from which to judge this complex situation.

I was one of a small group of behavioural scientists, five men and five women, a majority of university professors who were also management consultants, as was the leader and organiser of the visit, Dr. Donald C. Cole, of Cleveland, Ohio. Dr. Cole had been on a similar visit to South Africa several months earlier. He is also the founder and head of the International Registry of Organisational Development Professionals, an organisation with 3,000 members worldwide, who are committed to "nonviolent large systems change." I am not a member. Our "Behavioural Science Consulting Team," as it was known, went to South Africa with that objective in mind, hoping to learn enough possibly to suggest approaches to peaceful solutions to the present crisis. Obviously we were naively optimistic. But if the individuals in the group arrived in South Africa with any preconceptions about the situation there, their attitudes certainly included opposition to apartheid and support for black aspirations for greater racial justice and "freedom from oppression." There was one black woman in our group, and several South African blacks, including a high school principal, a high school science teacher, and the mayors of two black townships, who travelled with us some of the time. What follows, however, represents my own conclusions. In no sense did our group attempt to arrive to any agreed upon consensus.

We covered the country from the tip of the Cape of Good Hope in the south to the independent homeland of Venda in the north, close to the borders of Mozambique and Zimbabwe, visiting most of the major cities. We managed the usual tourist sites like Kruger National Park, but were primarily interested in the sociological and political situation, and in meeting South Africans of different races and political persuasions. We stayed not only in leading hotels, all of which were integrated but also spent several days with South African families, and were entertained by others. We spent one night with black children at a rural camp for disadvantaged urban youngsters, visited schools and universities, and on our first day in the country met in Soweto with striking black, high school students and several militant, black adults. We had interviews with cabinet ministers, with the heads of the prestigious Human Resources Research Council, and with the deacons of the six million member Zion African Church. Thanks to the good offices of my friend, Senator Claiborne Pell, now chairman of the Foreign Relations Committee, I, but not the others, had interviews with the then U.S. charge d'affaires - now the new ambassador, a black is in charge - and with our consul general in Cape Town, also a black. Both were opposed to the punitive measures then being considered in the Congress. I also met with the deputy editor of the leading English-language newspaper in Cape Town, which opposes the Botha government, attended services at the Anglican cathedral where Tutu was soon to be "enthroned" as Archbishop of Cape Town, and as did all of us, talked with many South Africans, black, white, and coloured. We were not the usual tourist group, but one of individuals hoping to learn what, if anything, might be done to ameliorate the present dangerous state of affairs and avoid further violence.

South Africa is a very troubled country and, in my opinion, more trouble and increasing violence is inevitable. I do not think, however, that the world is witnessing what Anthony Lewis in his syndicated column on 20 January called the "dying agony of white supremacy."

Let me now deal with the major issues of the current troubled situation; apartheid as national policy; the nature and control of the national government; conflicts and violence; the punitive measures taken by the United States; the dilemma of the Botha government and its reaction to the punitive measures; and, finally, solutions to South Africa's problems, if any.

On apartheid, I have indicated that that government has no intention of giving up the basic aspects of the concept, that is, the provisions for separate living arrangements for the four racial groups - the group areas, the black township, and the homelands - although modifications in some respects have already been made or promised. However, many of the restrictive provisions of apartheid designed to prevent contact among the races have been eliminated within the last two years. There has been a substantial "dismantling" of apartheid. Among significant government actions have been the scrapping of the Prohibition of Mixed Marriage Act of 1957, and of the sections of the Immorality Act of 1957, which prohibited sexual relations across the colour line. The strict segregation of places of public accommodation has largely been terminated, although President Botha has expressed opposition to compulsory integration. Most hotels, restaurants, clubs, and cinemas are integrated as are libraries, post offices, private hospitals, and parks. Recently public transportation was integrated, as is South Africa Airways, and even accommodations on the famous Blue Train. (It will be remembered that it is less than twenty-five years since public transportation in some southern states was desegregated). One extremely significant measure has been an amendment to the Group Areas Act that opens up free trading areas in the central business districts of cities, allowing members of all racial groups to acquire and occupy premises for business purposes. Another major step in dismantling apartheid has been the abolition of the hated pass laws, which inhibited the movement of both whites, who were not allowed to enter black townships without a permit, and nonwhites. In 1984, 194,000 persons had been arrested for violating these laws. Sports have long been integrated and now even beaches are being integrated, although not without opposition from some whites. The Anglican and other English-language churches have long been integrated. The largest white church in South Africa, the Afrikaner-language Dutch Reformed Church, has recently disavowed its long-standing Biblical justification for racial segregation. Although the individual white Reformed churches are not yet integrated, I suspect that with over a million blacks and some 700,000 coloured in its divisions for these races, integration will occur before long, in each case depending upon the wishes of the congregation. The largest black church is the Methodist, with over eleven million members.

Significant progress toward integration has been made in education, although primarily at the university level and in the private schools, most of which are church-related and teach in English. About 43,000 whites attend private schools, but blacks, coloured, and Indians are beginning to be enrolled. Public schools are still segregated. In 1984, there were 2,539 schools for whites with 981,648 pupils. There were 17,288 schools for blacks, with 5,796,221 pupils.

In 1985, there were 458 schools for Indians, with 233,660 pupils, and 2,062 schools for coloured, with 800,377 pupils. There is no doubt that inequality exists between white and black education. But as the 1986 *Information Digest* of the South African Foundation states, "No matter what resources are applied to black education in South Africa, a major limitation remains the shortage of qualified teachers." Progress is being made, however. In 1981, the *Digest* reports, the "total number of aspirant black teachers enrolled for qualifying courses was 436. In 1985, the figure is 7,300." But it concludes, " . . . equality in education remains an objective rather than a reality."

I believe that it will be a long time before the public schools are fully integrated, although I expect some earlier experimentation with integration on a token basis. It must be remembered that integration in the United States was ordered only in 1954; that federal troups had to enforce integration in Little Rock, Arkansas, in 1957; that attempts to integrate schools in Boston and other cities by busing resulted in substantial violence; and that the schools of San Jose, California, were integrated only last fall (1986) after a long series of court battles. Few Americans would argue, moreover, that even today schools in the black inner-city ghettoes are equal to the schools in the white suburbs.

When one examines higher education, one notes modest but commendable progress. The institutions of higher education have multiracial student bodies. The current system includes eleven originally all white universities, five employing the Afrikaner language, four the English language, and one a "parallel-medium" institution. These ten had a total enrollment in 1985 of 134,728 students. The University of South Africa (UNISA) is a "distance-teaching" institution, offering what we consider an extension program, including correspondence courses, and enrolling 77,028 students in 1985, two-fifths nonwhites. There are eight white "technikons," more like junior colleges, with an enrollment of 31,948 in 1985. In the four English-language universities, about 15 per cent of the students are nonwhite. Most black students attending white institutions are in these universities. In the Afrikaner-language universities, there is only a token black enrollment, but a few years ago there were no blacks enrolled.

There are five black state universities with a total enrollment of 19,231, of which 18,947 are black. There are three universities in the black "republics" of Transkei, Bophutatswana, and Venda - an impressive new institution which we visited - theoretically nonracial, enrolling 6,730 students, but as of now only a handful of them white. The University of Durban-Westville, originally exclusively for Indian students, out of a total student body of 12,000 today, now has approximately 400 blacks and 200 whites. In 1985, there were 2,632 Indian students at other residential universities.

The institution for coloured is the University of the Western Cape, which in 1985 had a total enrollment of 7,242, with 2,419 coloured at other residential universities.

It is obvious that higher education has been desegregated, but real integration has a long way to go. Much of the problem results from the poor quality and small number of black secondary school graduates, since few even of the graduates can meet university entrance requirements. Progress is being made, however, and can be expected to accelerate even faster, regardless of the outcome of political conflicts. In any case, once again one must keep in mind in judging the university system of South Africa, the abysmal record of integration at colleges and universities in the United States.

In general, in addition to specific formal actions to dismantle apartheid and improve the situation for blacks and other nonwhites, the government has tended to relax other aspects of apartheid which remain in place. It has accepted the permanence of a large number of blacks in white designated areas, recognised it is not feasible to assign all blacks to their respective tribal homelands; and it is considering granting dual citizenship to residents of the homelands.

Perhaps the most significant single proposal for eliminating the stranglehold of apartheid on the country was a now aborted plan for power sharing in Natal Province. This was developed by a conference of delegates from thirty-nine groups of blacks and whites of various political persuasions, though without official representation of the National Party and with a scornful boycott by the radical black African National Congress and the multiracial United Democratic Front. The proposal was for a merger of white-dominated Natal and the black homeland of KwaZulu, which is located in Natal. The merged area was to be ruled by a two-chamber parliament, one chamber based solely on "one-person-one-vote," which would result in a black government with a black prime minister, probably Zulu Chief Mangosuthu Butelezi; the other based on equal representation for blacks, Indians (who outnumber whites in Natal), English speakers, and Afrikaner speakers. Provisions were made for eventual black rule over the whole area. The optimists looked upon this revolutionary proposal as a blueprint for an eventual peaceful solution for the current explosive impasse in all of South Africa. Unfortunately, the proposal was rejected by the National Party leaders for Natal.

This proposal, in my opinion, gets at the real heart of the problem in South Africa - that of citizenship, although equal economic opportunity and alleviation of black poverty may ultimately be even more important. It is not so much the residential and working separateness of apartheid which fuels the bitter conflict in South Africa, as it is the denial of voting rights to blacks, which is taken to be symbolic of the denial of other basic democratic and human rights. Blacks can vote only in the homelands or in black townships and then only on local issues. Recent legislation has liberalised to some extend the voting rights of Indians and coloured.

The key issue has become what kind of a national government will prevail in South Africa. Will it be a government primarily if not exclusively of whites, which is the position of the die-hard conservatives; a government based on one-person-one-vote resulting in a government dominated wholly by blacks, which, regardless of public pronouncements of its leadership to the contrary, is the position of the African National Congress (ANC); or by a multiracial government still based on one-person-one-vote, but with guarantees for shared power representing all four racial groups in South Africa? Three-quarters of the blacks, the polls show, favour a multiracial government. The ANC, however, wants not only a black government, but a socialist, i.e., a Marxist state as well. Once committed to nonviolence, the ANC has become convinced it can achieve its goals of overturning the white government only by violence. The ANC has strong ties to the South African Communist Party - 13 of its 30-member Executive Council are reported to be members, although not the two leaders, Nelson Mandela and exiled Oliver Tambo. The ANC receives support from the Soviet Union and Mrs. Mandela has stated that "the Soviet Union is a torchbearer for all our hopes and aspirations."

Committed to violence, the ANC practices it systematically, primarily using militant, young, mostly poorly educated, unemployed black youths, plus guerrilla forces trained in camps located in neighbouring countries. Even Archbishop Tutu, though originally supporting nonviolence, has now advocated violence, even by his church. It should be noted that Tutu has far more standing in the international communities, especially in the United States, than in South Africa. One of my professorial colleagues on the trip has written that he didn't talk to one black who thought much of him. He is seen as preaching sacrifice but living in luxury. The mayor of Katlehong, a black township of 300,000, who was with us several times, says Tutu "does not reflect the feeling of the black majority."

The factions advocating peaceful negotiations and a shared multiracial government include the white Progressive Federal Party, with 27 members in the white parliament, and the moderate blacks, whose major spokesman is Chief Butelezi, not only the paramount chief of the six million Zulus, but also head of the largest black political party, Inkatha, which has a million dues-paying members.

The present government of the South African nation is primarily in the hands of the state president, currently P.W. Botha, his cabinet of ministers, and now three parliaments. Until 1984, there was only one parliament, all white, elected only by whites. Today known as the Assembly, it has 178 members. In 1984, the constitution was amended to provide a House of Representatives for the coloured, with 80 members, and a House of Delegates for the Indians, with 45 members. All M.P.'s are elected, as is the state president, for five years. There is as yet no parliament for the blacks. The constitution provides that the "control and administration of black affairs shall vest in the state president." In time a black parliament will come, although it is too early to tell on what basis, but in any case, it will be limited in its powers, as are the House of Representatives and the House of Delegates. These have responsibility for legislation affecting their own communities, such as education, welfare, housing, etc. but they also share in legislation affecting general affairs, such as foreign affairs, defense, and security. On such matters, if agreement cannot be reached, the issue is referred to a multiracial President's Council of 60 members, whose decision is final.

Each parliament has its own council of ministers. The president's cabinet is appointed by him, from among all the M.P.'s, and members have come from both the coloured and the Indian parliaments. Two black ministers without portfolio and two black deputy ministers are now in the Botha cabinet.

At the next level of government are the four provinces, now administered by councils with members drawn from all racial groups, but with control remaining primarily in white hands. It is at the local level, especially in black areas, although to a lesser extent in coloured and Indian areas, where major problems and resulting violence has occurred. The blacks began to get some control over their local affairs after 1971, but effective control remained largely in the hands of white Bantu Administrative Boards. Gradually, however, blacks did gain control over their local affairs through town and village councils authorised by the Black Local Authorities Act of 1982. First elections for the councils were held in late 1983, but the turnout was low in the 24 town and 5 village elections held at that time, averaging 21 per cent of eligible voters. The militant black organisations had urged a boycott of the elections. Mayors of the towns are elected by and from the council membership.

By 1985, the large towns had councils with full local government powers. There were also 80 village councils. But the councils have not for the most part fulfilled the hopes of the planners, or the expectations of the black residents. In many cases they proved unable to provide adequate municipal services. Dissatisfaction, fueled by the militants, turned into aggressive action against the councils and their members. Radical blacks considered council members as collaborators with the Botha government, enemies of the black cause, with the result that violence against them intensified. Many councillors were forced to resign, some were killed - the deputy mayor of Sharpeville was hacked to death on his own doorstep - many attacked by fellow blacks, their homes burned, their families threatened.

Few Americans realise how the situation has changed from earlier manifestations of trouble. The Sharpeville riots in 1960 and the Soweto riots in 1976 were prompted by black opposition to the whites. That, of course, still exists and can be expected to accelerate, as some of the recent bombings in urban areas demonstrate. However, the violence of the last few years has been more against "the system," against those blacks who are part of the system - the black councillors, black policemen in the security forces (nearly one-half of the public force of 46,000 are nonwhites), administrators and teachers in black schools. Any blacks who cooperate with the government - even those who in opposition to a militants' boycott, pay rent for government-owned housing - have their lives at risk. Intimidation of such blacks by the militants is destroying the advancement of black rights which the government expected by its changes in the policy of apartheid. But as the South African Foundation points out, the reforms already made create rising expectations for more such changes, and thus the dismantling which has occurred results in ever greater dissatisfaction and violence.

Since racial unrest broke out in September 1984, over 2,300 people have been killed in the two year plus period. By comparison, New York City had 1,309 homicides in the first ten months of 1986! Of those killed in South Africa in 1986, the majority were blacks killed by other blacks, 284 in the four months ending 30 June. In the four weeks between 10 July and 7 August, 65 blacks were killed by other blacks, compared to 11 killed by security force action. Chief Butelezi has said that what blacks do to other blacks is worse than what the security forces do.

The brutality of these killings seems not to have aroused much criticism from the West. Three-fifths of the 284 blacks killed earlier last year were killed by the "necklace," a method by which the victim is tortured by having an automobile tire filled with gasoline placed around his neck and set afire. Often the victim's hands have been hacked off to prevent any attempt to remove the tire. Frequently the crowds stand around cheering and family members are often made to watch. Within minutes, the victim is burned beyond recognition.

Mrs. Mandela, so admired in the United States for her spirited opposition to the government held up a box of matches in April 1986 and boasted, "With our boxes of matches and our necklaces, we shall liberate this country." Oliver Tambo, who has been officially received in Washington by Secretary of State George Schulz, unwisely, I believe, has refused to condemn the necklace.

The attacks on moderate blacks are not confined to individuals but extend to their properties or facilities sponsored by them. From September 1984 to July 1986, blacks were responsible for damaging or destroying 985 businesses, 46 churches, 1,272 schools, and 3,920 private homes, among the homes, that of the high school principal who travelled with our group.

I see no end to the violence. The black mayor of Daveyton, a township of 150,000, who also travelled with us, was the speaker at a graduation banquet for blacks receiving certificates for completing their course on community leadership. Though very critical of apartheid, he was even more critical of the radical blacks. "Jungle justice is meted out against opponents of the radicals in kangaroo courts," he said, "where the semiliterate youths dispense stone-age justice and unashamedly sjambok (i.e. whip) their own fathers and mothers." He predicted that if the radical organisations continue with their blueprints for action, the "silent majority" of moderate blacks in the townships, "will one day rise in a revolt which will make the present unrest look like a Sunday-school picnic." Reaction has already set in, with vigilante groups, largely middle class blacks, mounting a violent backlash against the young radicals. The conflict, however, is not so much between economically different classes, as between the young and their elders. It has turned into a generational struggle. But as the economy deteriorates further, the economic factor will become more important. South Africa has been enduring a serious recession for several years. The rand at forty-eight cents U.S. is about one-fourth of what it was two years ago. As sanctions and divestiture result in more blacks being thrown out of work, with their families facing hardship and starvation, the mass of moderate blacks will indeed rise up against the radical blacks who have advocated sanctions and called for sacrifices needed for such action. Starving blacks, however much they deplore apartheid, will demand retribution, and Mayor Boya's dire prediction of catastrophe will be realised. Undoubtedly, age-old tribal animosities will surface as scapegoats are sought and exacerbate the violence of moderates against radical blacks.

Unfortunately, the imposition of sanctions by the U.S. Congress and other governments which follow the lead of the United States will hasten the catastrophe and contribute to its intensity. President Reagan was right when he opposed sanctions, which, he said on 9 September 1985, "would injure the very people we are trying to help." No one, white or black, whom we talked to in South Africa was in favour of sanctions. Obviously, we did not talk to Tutu, who on 2 April 1986, called for "comprehensive sanctions," instead of the selective sanctions which had been voted. Our traveler friend, Mayor Boya, in pleading with us to oppose sanctions, wrote that it was "very clear and undebatable that sanctions would cost the country millions of jobs." He also pointed out sanctions would result in most of the blacks from outside South Africa who are working in the country losing their jobs, with disastrous results on the already weak economies particularly of Mozambique, Lesotho, and Swaziland. In 1981, the Carnegie-supported Study Commission on U.S. Policy Toward Southern Africa, with four blacks, including the chairman, out of eleven members, opposed "disinvestment and other major sanctions" against South Africa. If that was sound advice six years ago, how much more reasonable now that so much progress has been made in dismantling apartheid!

The objections to divestment were just as unanimous among both blacks and whites. Chief Butelezi, in accepting an honorary degree in March 1986 at Boston University, stated that his party, Inkatha, every year at its annual general conference had unanimously rejected disinvestment as a strategy. It should be pointed out, incidentally, that while American firms leaving South Africa usually attribute the move at least partly to opposition to the nation's racial policies, the major reasons are economic, as General Motors, for example, admitted. As of October 1986, there were still 248 U.S. companies doing business in South Africa, but 63 had left since 1984. Others will go, but more because of what IBM termed the "hassle" factor operating in the United States, than from any determined opposition to policies of the government.

From the standpoint of an American corporate presence in the country, it would make more sense in terms of aiding blacks, to increase investment in South Africa, not pull out, thus creating additional jobs and helping to relieve the economic plight of increasing numbers of South Africans. Such investment has been urged by the black Episcopal bishop of Washington, D.C. His advice should be heeded.

The punitive measures have been counter-productive. The result has been not only a halt to further dismantling of apartheid, but also a toughening stand against reform. The imposition late last fall of heavier censorship of the press and television is a recent example, as is the jettisoning of the proposal for a multiracial provincial government in Natal. The Afrikaner is a stubborn individual, as the great Voortrekker exodus demonstrated. But even many of the more tolerant English-speaking South Africans are reacting negatively to the attacks on their country as a pariah among nations. Some are planning to leave, but most of them do not have British passports, and as several told me, are stuck there just as are the Afrikaners. The fact is that the white race in South Africa has nowhere to go.

External actions against the Botha government may result in another outcome which would have the opposite effect upon the black population than was intended. These actions carry the threat of the overthrow of the current government by a more conservative white government, which not only would end further efforts to improve the situation for nonwhites, but would undoubtedly see the reinstating of some of the worst provisions of apartheid which have already been repealed. That in turn could alienate the moderate blacks, strengthen the hand of the radicals, and result in bloody confrontations of blacks and whites.

At the time of the vote for sanctions, the Botha government was walking a dangerous tightrope, caught between the black radicals of the left, committed to overturning the government by violence, and the white radicals of the right, primarily conservative Afrikaners, who are determined to oppose any sharing of power with nonwhites, wanting to preserve white hegemony at any cost. Before sanctions, Botha was planning an even more significant relaxation of apartheid and greater accommodation of black aspirations. In the best book I've seen on the current situation, *South Africa - No Easy Path to Peace* (Century Hutchinson, 1986), Gordon Leach, a BBC correspondent in the country, writes that there has been a real change of viewpoint on the part of the government leadership: an admission that its policies "have been so misguided, have caused so much hardship to people, (that) they were desperately in need of revision" (p. 249). Since this was written, the measures taken against the country may well have caused a reversal of this more compromising view.

Certainly it was clear when we were in South Africa that if Botha went too far in meeting black demands, he ran the risk of a revolt in his own party which would result in a more conservative party in power. The danger was obvious. A new party, The Conservative

Party, had 19 members in parliament, but only two had been elected as party members. The rest had merely changed allegiance. And in the wings is the ultra-right, neo-Nazi Herstigte National Party, which though having only one M.P. had received 13 per cent of the white vote in the election of 1981. There will be new elections this spring. What the outcome will be is at this writing (January 1987) unpredictable.[1] But there can be no doubt that the best interests of the nation, including the blacks, lie in the re-election of the Botha government. With a substantial victory, the president could begin negotiations with all factions, including the African National Congress. Botha has indicated that he is willing to sit down with the ANC leaders if they renounce violence. This they have thus far refused to do. But a compromise on both sides may be possible.

Is there a solution to this troubled country? If so, can it be found before current violence escalates into a real bloodbath? Obviously, no one can be sure, certainly not an outlander like me, with only a limited knowledge of the situation. But I believe that there is no way the white minority in the near future will accept a wholly black government, especially a Marxist one. The situation in Mozambique is enough to convince even the most liberal of whites that this would be disastrous for them and for South Africa. Memories of Idi Amin in Uganda and of Bokassa in the Central African Republic, and even the current situation in neighbouring Zimbabwe, strengthen the determination to avoid at all costs, even bloody civil war, the installation of an all black government, with the tribal enmities prevalent in the background. Leach writes that "The whites look at the tribal makeup of their own country and firmly believe that what happened in the rest of Africa once the blacks gained power would pale into insignificance compared to the tribal wars which would be unleashed in South Africa if the blacks took over" (*op. cit.*, p. 120).

On the other hand, there is no doubt that even the moderate blacks will not much longer accept a government in which they have so little participation. Mayor Boya predicted that the longer the government delays in "sharing the power and wealth . . . the stronger the chances of its violent overthrow increases." We were told by one of the militant black leaders in Soweto that if the blacks fail to "force

---

[1] In the spring election Botha and his National Party were returned to power by a comfortable margin, but the Conservative Party made significant gains and became the official opposition (editorial comment).

capitulation" to their demands, their plans call for killing white women and children. An exiled black journalist, Dumisani Kumalo, an ANC member declared at a symposium on South Africa at Middlebury College last spring: "South Africa will be free. It might take two days, it might take twenty years. There is no way four million whites can continue to oppose 26 million black people." I think history demonstrates otherwise, if the overlords are not held back by humanitarian considerations. But what Mr. Kumalo failed to point out was that in less than twenty years, by the year 2000 the blacks will number 50 million, if the present birthrate continues. This is what Leach calls South Africa's "time bomb." There is little hope of stemming this black tide, nor of the government's providing within a dozen years the housing, education, employment, welfare services, etc. for the expanded population, thus aggravating today's explosive situation and leading to incredible levels of violence.

There will certainly be an escalation of today's violence and killing, but I do not believe the government can be overthrown by violence. Outnumbered though they are, the government forces will prevail if the situation comes to a civil war. The fire power lies with the whites. South Africa may have the best trained security forces in the world. They undoubtedly have or could produce small nuclear weapons and in a showdown might be tempted to use them. Soviet equipment, if it were to be provided to antigovernment forces, and Cuban mercenaries, if sent in, could delay but not determine the outcome. Hopefully, however, before any struggle would reach such a phase, both sides would be willing to sit down and eventually work out some sort of a solution which would partially satisfy each side and end the conflict. The Carnegie Commission concluded in 1981: "There are no easy solutions for South Africa. The choice is not between 'slow peaceful change' and 'quick violent change' but between a slow, uneven, sporadically violent evolutionary process and a slow but much more violent descent into civil war." Both paths, it states, "could lead to genuine power sharing."

This is still true, whether sooner or later, there must be a multiracial government. The whites may not be in charge, as now, but their security and their protection must somehow be guaranteed. Eventually, the climate for peaceful existence in South Africa will be achieved. Two developments, I believe, will be especially important in creating this climate. One, creating a strong black middle class, the beginnings of which are already in place, and to which divestiture may actually contribute by providing more skilled, managerial, and entrepreneurial opportunities for blacks. The second is the upgrading of the education of blacks, and the gradual integration of the schools.

As black schools are improved, more blacks will be able to attain university training. This will create more upward economic mobility, thus contributing to the goal of a more substantial middle class. Moreover, as integration proceeds, children who go to school together will learn to live in relative peace and harmony, although as the experience with school integration in the United States demonstrates, complete racial peace is probably unattainable.

With time, tribal loyalties will be reduced, as they are now in the urban areas. Eventually, the racial segregation required by the Group Areas Act will be abolished, regardless of the Botha government's statement to the contrary. The homeland effort will be downplayed - it has already been curtailed - and as the homelands become more secure economically and more viable politically, they will no longer seem to be undesirable living arrangements.

All these changes will take time, patience, understanding, and commitment. But the world is impatient. Americans especially expect this to occur almost overnight, and seem to believe that our punitive measures will speed up the process. They can no more do this than the earlier ambiguous policy of "constructive engagement." Only the South Africans can determine their own destiny, and as I have suggested, increased pressure tactics will only be counter productive.

Have we Americans forgotten the long effort to achieve racial equality in this country, still so short of achievement? Jesse Jackson tells us that he "grew up in a system of apartheid in South Carolina." In fact, there is evidence that racial equality is actually retrogressing with a resurgence of racial hatred in the United States. Apartheid is certainly a bad "abstract idea;" indeed it is undoubtedly "immoral," as so many American critics allege. But are we justified in pointing the finger of shame at South Africa, given our own dismal record of coping with a much less serious problem in this country? The bloody Civil War was fought 125 years ago over the issue of racial freedom, if not racial equality. An attempt to legislate the latter nationally occurred only in 1964 with the Civil Rights Act. But racial incidents occur somewhere in the United States almost daily, and not only in Forsyth County, Georgia. Even on university campuses, where the dramatic opposition to apartheid mushroomed, four out of five black students at predominantly white schools state that they have been subject to harassment because they are black.

Much of the condemnation by Americans of South Africa's racial policies, and I do not support or condone those policies, stems, I believe, because it relieves our guilt feelings over our treatment of blacks - and American Indians - from our chagrin that we have failed to fulfill the ideals of racial equality to which we play lip service - consider the virtual canonisation of Martin Luther King in connection with birthday celebrations around the nation! So we vote for sanctions because, as Senator Robert Dole commented after the override of the President's veto, "It made us feel good." And while I am no admirer of arch conservative Senator Jesse Helms of North Carolina, I believe he was correct when he stated that the sanction vote "was purely an election-year trauma . . . a purely domestic affair." He suggested that the 100th Congress should repeal sanctions - which it should. It should also urge American reinvestment in South Africa.

Instead of "moral posturing and futile gestures," the *Providence Journal-Bulletin* correctly editorialised on 23 November 1986, we should provide "long-term practical assistance to moderates of both races." If the leaders and planners of the National Party in South Africa based their government policy for the last forty years on an abstract idea which was flawed from the beginning and therefore eventually doomed to failure, an idea which defied a permanent solution which would be sound, just, and feasible in today's world, the United States government and the many Americans who support its position on South Africa, are also at fault for having set a policy which is not sound, nor just, nor feasible, nor capable of realistic implementation.

PART II

APPLIED GEOGRAPHY AND PLANNING

CHAPTER 5

REFLECTIONS ON GEOGRAPHY AND PLANNING

T. Walter Freeman
Emeritus Professor of Geography
University of Manchester
Manchester, England

INTRODUCTION

Some geographers, perhaps just those who are now, as the Anglican prayer book expresses it, of "riper years," were taught to think of the world on three planes, global, regional and local. They had an abiding curiosity about distant as well as near countries and places, about regional areas of appreciable size such as the Scottish Highlands or New England, and of small areas where a lot could be seen on foot or by some relatively slow means of communication such as a bicycle. This paper is one geographer's meditation on man and the land.

No attempt is made to assess the range of work done at the meetings of the Applied Geography Commission of the International Geographical Union with which Peter Nash was so closely associated. At those meetings one had the opportunity of meeting geographers from many countries, and in the papers given as well as in conversation "in my country" became a familiar phrase. Was India as one had long imagined it to be? What contribution had geographers made to the development of Siberia? How had Sudentenlands of Czechoslovakia fared during the post-war period? Exactly how could the growing world impact of tourism be assessed? Did the definition of administrative areas of varying extent bear a reasonable relation to the distribution of population? All these are questions of social and economic significance on which geographers might be expected to have something useful to say, but the value of their comment must rest on a basis of geographical understanding. As Eva Taylor in Britain said, "It is no use talking of applied geography unless one has a pure geography to apply." An attempt is made in this paper to ask questions rather than to answer them and for that no apology is offered.

A GENERAL VIEW

What can we know of the future? From time immemorial people have wanted to foresee events, to look forward with hope or perhaps with gloomy foreboding, possibly to learn from the past, but all who are mature in years know that the future is veiled from our eyes, mercifully perhaps, and on balance both worse and better than what seemed likely to happen. The success of Ecology parties (recently dubbed "the loony green left") in some countries shows the mixture of fear and hope on the future of the environment that is characteristic of many thoughtful people. On the one hand convinced that mankind is using the resources of the world at a rate unprecedented in human history (a view held by many responsible people for a long time), they may also have the hope that with modern scientific knowledge the needs of people for a reasonable, and indeed rising, standard of living can be met. At one time it seemed that nuclear energy was to be the solution of the energy crisis but now its use seems to be fraught with dangers. This situation is demonstrated when driving along the Ml motorway in England when one sees a sign "South Yorkshire nuclear-free zone." This area, around Sheffield, is often called the "Socialist Republic of South Yorkshire." And, it is not the only "nuclear-free zone."

Maybe our fears are our worst enemies. In the difficult but fascinating years following World War II, as indeed for earlier times including those of Malthus, there was the fear that the world's population was increasing at such a rate that it would no longer be possible to provide food for the growing multitudes. Dudley Stamp and many more people of responsible concern held this view, but Stamp eventually came to the decision that the farm lands of the world could feed its population and that the crisis lay in distribution rather than in agricultural limitations. This was not a new idea for it had been well demonstrated in a now forgotten classic, W.H. Mallory's *China: Land of Famine* (1926). Mallory showed that there could be a surplus of rice and other basic foodstuffs in one part of China while hundreds of thousands were starving elsewhere, but neither the transport nor the organisation to combat the famine with effective action existed within China. That intensive agriculture is commonly found near towns has long been a commonplace, however, the apparent miracle of this century is that it seemed to be scientifically possible to make two blades of grass grow where one grew before. Obviously, immense questions were raised by the use of the vast resources of the U.S.S.R. beyond the Urals during World War II. But those who have visited areas of recent agricultural development realise that there are still "limits of land settlement," to use a once famous phrase; e.g. as seen by those who look carefully at irrigated areas and notice an outer fringe marked by salinity. India's Green Revolution is hailed as a triumph but that does not imply that all of India's problems of poverty and malnutrition have been solved. Nor can one be assured that China's limitation of families to a single offspring can meet the present fear of actual or potential overpopulation.

Naturally those who care about humanity in general see much that is frightening. There is clearly no need to deal here with the possibility of world destruction through political madness, of events more terrible than anything ever known in human history: that is common knowledge. Nor has the author any intention of luxuriating in an outburst of moral indignation: that can be left to the bleeding hearts brigade. We all wish to see a better world. The tragedies of two world wars in this century have been deeply engraved on the hearts and minds of multitudes and yet out of all the suffering the human spirit has shown a resilience that has resulted in better living conditions for a substantial proportion of the world's population. It may still be true that the rich are getting richer and the poor poorer but at least with modern means of visual communication, especially by television, the problems are made known and through various international organisations, some depending on private charity, help may be given, however inadequately. And there are now hopeful signs that in Western Europe governments will respond more quickly. One basic problem is that in some areas relief agencies are hampered by civil wars or other intractable circumstances. Those who work for Oxfam or Christian Aid find that the response is more widespread than formerly and that this situation may well be due in part to television programs which have brought the realities of African famines into millions of homes.

Do we still want to know something of the whole world? When the present author was a young geographer one was expected to do so, and in many universities A.J. Herbertson's "natural regions," in effect climatic regions, were often used as a basis for first year courses. One embarked later on the various schemes advocated for more detailed recognition of regions (perhaps encountering Fenneman on one's academic way) which were aware of physical features, mountains, uplands, lowlands, plains, karsts and the *pays* of France as a shining correlation between distinct physical landscapes and agricultural practices. Some of the authors of such regional schemes based on the countryside as the prevailing scene seemed rather embarrassed when they came to great metropolitan areas such as London or Paris: were major cities themselves regions? As a substantial proportion of the world's professional geographers were working in Europe, with its markedly varied landscapes, regional geography had its compelling appeal. In aspirations, if not perhaps in realisation, it dealt with the whole of the environment, however used; and it was historically minded. Some of its advocates explained that an evolutionary outlook should prevail, for the landscape was an end product developed from the first beginnings of human settlement.

Steadily a new industrial and commercial population was imposed on the agricultural landscape, located in towns and industrial villages such as those around mills and mines. There were of course towns long before the Industrial Revolution, for convenience regarded, despite the misgivings of some historians, as beginning in the middle of the eighteenth century and gradually spreading through the entire world, but the general effect was to make a new balance of population between "urban" and "rural" and in time to make urban living normal for a majority of the population in many countries. The countryside was an area to be "visited" rather than experienced and the social environment became more immediately real than the physical environment. It was the people on the land who were in intimate touch with the physical environment and in many geographers' minds there was a division between work concerned in some way, however indirect, with industrial production and that of food production. In fact some people saw beyond such a false dichotomy for even in the middle of the nineteenth century some perspicacious writers were saying that "Agriculture is itself an industry." Now it is so to an infinitely greater extent. The inevitable question is whether it is an industry capable of destroying the earth. Turning again to the Ecology Parties, with their reservations about the use of chemical fertilisers, are we sure that their warnings are pointless? Does the removal of hedges and of woods presage severe problems for future generations? And are the current developments in Amazonia wise or, as some fear, scientifically suicidal, leaving aside the massive human problems involved?

In short the geographer is driven back to basic problems and especially to those of landscape characteristics, to the old questions of man and environment. As one travels west from Moscow, the land is spaced out in vast fields with the buildings of the collective farms visible at intervals, as well as the small but highly productive allotments of individual householders, invariably having space for a few bright flowers. Some of the old farm houses survive, with multiple occupation suggested by as many as six separate television aerials. Cross the border into Poland, and although the official guide will decisively explain that collective farming has inaugurated a new and better age, the small farm is still widely present in the landscape. Germany has fine farms and every sign of efficiency. In the Netherlands space is the scarce commodity, for the houses and gardens are small, the fields intensively cultivated, with control of water as the inevitable basis of farming and indeed of urban living in a land having substantial areas reclaimed from the sea.

On two of the field excursions of the Applied Geography Commission the agricultural landscape scene was clearly an expression of public policy. In Czechoslovakia the large fields and collective farms are dominant and Professor Scheidl from Vienna explained that the landscape of Bohemia had been changed out of all recognition since the time when as a young man and he cycled round its roads and lanes with their innumerable small farms. Later, in Brittany, the host geographers showed the effects of *planification*, which included the bringing of new industry to country towns as well as to larger centres and the replacement of the old *bocage* farming with its abundant hedges by large modern fields bounded by electric fences for intensive dairying. Aesthetically the changes are not pleasing, but people cannot be expected to live as picturesque relics in a landscape that is underfarmed and less productive than it might be with such amenities as electricity and piped water. Romanticisation of "peasant culture" is a luxury enjoyed by those who are not in any danger of experiencing it on a day-to-day basis.

Every landscape reflects the experience and desires of its people, subject in many countries to governmental control. People are far more interested in their home environment than is generally realised: some are obsessed by it. Anyone who has lived in an English village or town will know that there are many residents resentful of change, eagerly guarding the past through the retention of old buildings and of open spaces. David Lowenthal has written eloquently and engagingly on the "nostalgia" so prevalent in America and Britain. A visitor to New England will certainly be charmed by the taste and care lavished on the preservation of eighteenth and early nineteenth century "colonial" houses and on the public buildings such as the splendid churches built in central Boston. He may well be amused by the efforts to create the "simple life" as that enjoyed by Marie Antionette and her ladies in the "village" within the palace grounds of Versailles.

Even so, he may also reflect that the wishes of the privileged aristocracy have gradually spread through society. Those who work in cities may have a "second home" as a retreat from urban conditions, used for holidays and weekend or other short visits and perhaps as the eventual retirement home. It is not the universal wish for many city workers are content to live in suburbs or pleasant towns within daily reach by trains or other means of transport. In some parts of Britain, however, social problems have been caused by the demand for "second homes," as with a free market in house purchase, people from the cities can outbid local people wanting homes. Indeed in Wales there have been occasional attacks on such "second homes" by extreme nationalists who object to any dilution of their own culture by non-Welsh speakers and have even been known to resent the export of water to English cities, notably Birmingham. Climatically, it would seem, Wales is over-abundantly supplied with water. Are such fears irrational? One remembers with interest a visit to the area of Sweden close the Sound, at a time when there were various schemes for building a road bridge during which it was explained that one unfortunate effect could be a sudden demand from people living in Greater Copenhagen for existing properties or land to build new holiday homes in an essentially rural area with fine beaches along the coast.

It is not surprising that one "growth area" in geography since World War II has been the study of recreation. Much of this depends on the increasing mobility provided by the motor car, clearly something people want. Many of those who crowd the roads to cities on Sunday evenings have been enjoying "a day in the country," varied in form from the athleticism of mountain climbing to a gentle progress with the children round a "wild life" or "safari" park. Modern motorways have brought quite distant areas within reach for a day visit, for example, from the Manchester area a journey by car that begins about 7 a.m. can make possible a walk of perhaps eight hours on the mountains of the Lake District or Snowdonia national parks. The attitudes of people are always a fascinating study and such large numbers are drawn to particular mountains that some paths are now eroded and in places dangerous while other mountains, of equal scenic interest, are visited by comparatively few people. Do people behave like sheep? Naturally every experienced mountain walker knows that there are dangers in isolation and some associations, such as Cairngorms Club in Scotland, make a rule that there must be at least three people in each party so that in the case of accident one may stay with the injured member and one go to summon assistance. Splendid work is done by the mountain rescue teams in Britain though it is a sad commentary on modern life that some of their stores of equipment in the mountains have been vandalised.

Where do we seek some form of reality as geographers? Assuming that we have some respect for the old phrase "man and the land" the answer must lie in observation of the living world as one line of approach. We may - indeed must - gather statistics, work out densities of population, note its increase or decrease in various areas, consider the possibilities of transport by canal, road, rail and air, assess the consequent movement as one of the basic patterns of existence, map the use of the land whether for agriculture, forestry or some form of open countryside, calculate the intensity of settlement and possibly the incidence of crime, disease and poverty in towns as well as their commercial, industrial, recreational and other facilities: the possibilities are endless. But many geographers who use such sources will be constantly asking themselves how such data are reflected in the landscape. To take just one example of an invalid jump in reasoning: in some countries, as in Britain, the average number of persons per household has declined so that a town with a stationary or even decreasing population may require more dwellings for its residents. On the ground, new or recent building would suggest an increase of population, which is not in fact the case. Housing experts have various ways of defining "overcrowding" (beyond the scope of this paper), and expectations of space in dwellings differs from one family to another according to income, experience and even government policy. Visitors to the U.S.S.R. are likely to maintain a tactful silence when the Intourist guide explains proudly that through the beneficient government she has a three room apartment for herself, her husband, two children and a resident grandmother.

Probably a high proportion of the people who have become professional geographers derived part of their initial inspiration from their own home area (perhaps with the encouragement of some teacher): whatever its qualities any district has its own individuality. Nevertheless, a landscape different from that known in childhood may have a compelling appeal as in the case of Dudley Stamp who, though a townsman in origin, identified himself with the rural lands of the earth and combined in his work an understanding of its physical, scenic and economic character. Undoubtedly some areas are more interesting than others and those who live in areas of considerable variety are better placed than those located where there is little change of scene over scores or even hundreds of miles.

To a European the air journey from New York to Arizona has the fascination of revealing that the Appalachians really are like the topographical maps studied during geomorphology courses, having relief features that challenged the workers of the famous United States Geologic Survey who gave so much to geomorphology as well as to geology, but once beyond these mountains the apparently endless cultivated plains are something unfamiliar in experience. There are of course areas of comparable uniformity in the Old World. During a day in Sweden on a field tour of Germany and Denmark from Manchester University, Professor Torsten Hägerstrand was asked to have the bus driven for a few miles into the forest so that the point could be made that such forest, interspersed with clearings of varying extent, was the characteristic landscape all the way to the Pacific Ocean. Would such uniformity suggest that the taiga belt could not inspire good geographers? By no means, for Sweden and Finland have been distinguished for their excellent contribution to geography and among the Russians, also having landscapes of vast apparent uniformity, one major contribution has been a close analytical study of soils. The merit of a geographer as an observer lies in his ability to see the essential quality of a landscape, to discern its problems and to ask fruitful questions. Nowhere need be "dull." Sadly, long experience of conducting field tours for students, confirmed by conversations with others who have done similar work, leads to the view that a party will have three groups of observers, those who will see a very great deal, those who see what is within their normal range of interest - probably in the life of towns and villages or what is thought to be "special" because it is mentioned in a guidebook - but with only a modicum of general curiosity, and those who might just as well have stayed at home. But no party is more rewarding to a geographer than one composed of professional geographers.

It might seem that geographers have always regarded direct observations as a major source of light. Historically, the interest in geography developed from the records of travellers to new and previously unexplored areas different from the homeland, accompanied by route maps and possibly sketches, sections or other illustrations. According to Guyot, Carl Ritter once said, "What new information could I derive from a visit to Palestine? I know every corner of it." And he probably did, in sofar as it was revealed from the travels of others. His own work on Palestine filled several volumes of his massive *Erdkunde* (1822-1859), later condensed, translated and adapted for biblical studies by W.L. Gage and published in four volumes as *The Comparative Geography of Palestine and Siniatic Peninsula* (1866). But could one now believe that his work was of the same calibre as George Adam Smith's *Historical Geography of the Holy Land*, based on biblical scholarship and his own journeys, first published in 1894 and reissued in some twenty-five editions since then?

MAJOR PLANNING PROBLEMS: THE AMERICAN AND BRITISH EXPERIENCE

Four months at the University of Rhode Island in 1967, following the visit during 1966 for a meeting of the Applied Geography Commission, were a revelation of the changes that may come in a landscape through the course of a country's economic history. Even on the campus itself one could observe the old walls of fields long since abandoned and now submerged in woodland: they are also a familiar feature of the countryside. If ever there was an example of Isaiah Bowman's saying, "Man takes the best and lets the rest go" here it was. In time one learned that only one per cent of the state of Rhode Island was used for agricultural purposes and to a visiting Englishman coming from a country where land is a precious possession to be used to its maximum capacity as a matter of social obligation this was a revelation even more startling than anything recorded in Jean Gottmann's *Megalopolis* (1961).

Though well aware that much has changed in the United States since 1961, the author cannot resist quoting extracts from the review he wrote of Gottmann's book at the time:

> Six hundred miles long, from Boston to Washington, and 30-100 miles wide, Megalopolis covers 53,575 square miles, less than 2 per cent of the area of the United States, and in 1950 had 32 million people, over one-fifth of the total population. (It) is slightly larger than England, which in 1961 had a population of 43,431,000 . . . Megalopolis is the "most advanced case of an urban region (p. 215) because in and around Megalopolis a large degree of *de facto* suburbanisation has been achieved in areas still considered nonurban" (p. 39). The American census category of "rural nonfarm population" has been extremely useful . . . a map on p. 36 shows that in almost every county of Megalopolis it is at least 60 per cent of the whole rural population, and at least 70 per cent even in places over 90 per cent of the whole, on the entire northeast seaboard . . . The tilled area, having reached its maximum in 1850, has been declining ever since; half the area of Megalopolis is woodland, much of it abandoned farmland . . . . And although there is some highly intensive farming, many farmers keep part of their land for sale to home builders and many more hope for the biggest harvest of all - sale to the advancing city. Four things have made this possible: the

> excellent roads, the prevalence of cars, the rapid expansion of retail shops in the countryside and the growing disadvantages of life in city apartment and suburban areas. Essentially it is a two cars to the family civilisation . . . Is Megalopolis an awful warning to western European countries? . . . "It can't happen here" - or can it? (Freeman, 1962, pp. 534-35)

To a European visitor it would seem that in the United States money can buy anything, including land for house plots so that there is an indiscriminate spread of housing along the main roads and of varied enterprises along the repellent "automobile rows" near built-up areas. Some detailed information is given in Gottmann's chapter on "The Symbiosis of Urban and Rural" which opens with the comment that:

> In Megalopolis in the 1950s the interpenetration of urban and rural had achieved a complexity and a size yet unknown anywhere else on the globe. . . One may . . . compare the process observed in Megalopolis to what has taken place in the most urbanised parts of northwest Europe: in England, around London and in the Midlands especially, in the Greater Paris area of France, in the Ruhr industrial basin of Germany. Each of these regions is much smaller in area and more densely populated than Megalopolis. (Gottmann, 1961, pp. 217-21)

In fact Megalopolis epitomises exactly what planners in Britain and in continental Europe wish to prevent, for the precious commodity in such areas is land, and it seemed strange to the American students with whom these issues were discussed that the "house plot" in Britain should be "so small." They may not fully appreciate that the areas around the houses in Britain are normally carefully cultivated gardens, ideally having flower beds, even vegetable plots in some cases, and an area of lawn (or perhaps it would be more realistic to say "grass") variously used as a play space for children and a leisure area for adults. In practice, what one commands in living conditions depends largely on income. Generally blocks of flats (apartments) are appreciated only by certain elements in the population such as single people, childless couples or the elderly no longer wishing to cultivate a garden though glad to have access to some form of communal open space.

In Britain much of the propaganda on slum clearance, the provision of better housing for the artisan community and others of limited resources, dates back to the apostolic vision of Sir Patrick Geddes (1854-1932) and to advocates of the "garden city," such as Ebenezer Howard (1850-1928), whose work on model town planning was finally recognised by the passing of the Town and Country Planning Act of 1947. Some major firms such as the Cadburys in Bourneville, Birmingham, and the Levers of Port Sunlight, Merseyside, built houses and financed a variety of social and athletic amenities for their employees. There were many more such enterprises, some more successful than others. But in general the building of houses on a profit-making basis had resulted in slums, not only in the major cities but in smaller towns as well, and after the 1914-1918 war the first priority was to meet an accumulated housing shortage ("Homes fit for heroes to live in," boomed Mr. Lloyd George) and after that to replace the slum houses with more hygenic homes. The result was that between 1919 and 1939 there was a vast urban expansion, both of houses to rent from municipalities and of privately owned dwellings in favoured suburbs. After the 1939-1945 war a comparable shortage of houses was admitted but the solution was sought in part by building "high-rise" blocks of flats comparable with those in continental European - and perhaps especially Scandinavian - cities, for there was a wish to maintain definite social communities in the cities to avoid the creation of an urban "wasteland." Regrettably some - not all - of the dwellings provided on several floors are not now favoured for a multitude of reasons, many of which have not so far been convincingly explained. At present the investigations of Alice Coleman, from 1960 Director of the Second Land Utilisation Survey of Britain, and from 1978 Head of the School of Human Environmental Studies at King's College, University of London, and her colleagues promise a more adequate explanation than that so far given with all sincerity and sympathy by a variety of people. These problems are discussed in Coleman and Power, "Trouble in Utopia" (1984), and in Coleman's *Utopia on Trial: Vision and Reality in Planned Housing* (1980).

On one point in Britain there is general agreement. However hard to achieve, there must be effective preservation of the Green Belts around major cities, a definite "urban fence" dividing rural and urban areas, possibly with New Towns or planned extensions of existing towns as overspill areas from the great cities. This policy in general terms has received approval for it is clear that nobody wishes to see the extinction of the rich agricultural belt that lies between the conurbations based on Merseyside and Manchester, or the unlimited spread of settlement between Manchester and the West Midlands conurbations. On a more local scale the same outlook applies: from the centre of Birmingham in the West Midlands conurbation to the centre of Coventry it is just eighteen miles but the intervening (narrow) Green Belt is carefully preserved. But the problem is perhaps most acute in the southeast, much the most prosperous area of Britain at the present time as for a long time past and any foreseeable future. The immense demand for houses has driven prices to a level about three times that in the less fortunate areas of the country and to fantastic levels (the word is deliberately chosen) in London. Some firms in the southwest find that it is difficult, even impossible, to find people with appropriate skills locally and hope to attract workers from areas of high unemployment, such as Liverpool or Newcastle-on-Tyne. They fail to do so because the people from such areas can neither sell their houses at a reasonable price nor afford prices prevalent in the southeast.

Naturally such problems are acute in periods of high unemployment, such as the present time, and as always the actual incidence of unemployment differs widely from one area to another; clearly there is no easy solution available. Housing in Britain is subject to normal market forces, to the law of supply and demand in an economic situation that is difficult to predict. Whether the government should assume more responsibility for the changing location of industry is a matter of political debate into which it is not proposed to enter here. But it would be unfair to leave this matter without commenting that among geographers in Britain there are many devoted workers who are concerned with such problems as unemployment, social welfare, health, public transport, recreation, and many more aspects of urban living, including provision for people with special problems due to age, infirmity or disabilities. Perhaps it has taken the shock of riots in various cities to awaken people to the malaise of many communities. Whatever the reason, it is abundantly demonstrated that geographers are deeply aware of the social environment.

One major result of British planning legislation has been that extensions to existing settlements, both in villages and small towns, are in general aesthetically tolerable. Instead of the ribbon development along major roads that was so prevalent during the interwar years (as for centuries before then), many such places have a number of quiet roads with modern houses, all with gardens and garages varying in type according to income level. The secluded estate on which there is no traffic except that generated by the residents has become favoured. In particular many villages have received, if not necessarily welcomed, new residents working in neighbouring towns. Widespread in England is the annual "best kept village" competition organised by Rural Community Councils. The judges are not expected to have any interviews with the local people or to enter any premises; rather they look closely at the car park, garden, and external appearance of the village public houses (inns) but not its interior. What is of interest is the state of upkeep of the parish church and its graveyard (but not the architectural relics of a thousand years that may be found inside), the playing fields, public hall, war memorial, gardens of houses and the external upkeep of the houses themselves, grass verges if they exist, notice boards commonly located outside the parish church and elsewhere or in the window of the post office, shop windows, bus stops (whether litter free) and even telephone kiosks (many vandalised). Generally in the author's experience the standard of upkeep is high though it was not conspicuously so in the case of one village which had a notice saying "Great clean up: May 10, for the best kept village competition: all welcome." Judging began on May 1 and he with a colleague was there on May 6. No doubt this enterprise is in line with what David Lowenthal, in collaboration with Hugh C. Prince, wrote in "English Landscape Tastes" (1965) and later articles. They are well aware of English eccentricity, expressed in a liking for a landscape in which:

> Seemliness and propriety are respected; untidiness, however prevalent, is felt to be ill-mannered and offensive; demarcations are clearcut. Neatness is a matter of boundaries as well as of areas. Roadside verges, hedgerows, fences and railroad rights-of-way are trim, discreet, unambiguous . . . The scraggly growths of briar and bramble, the motley ground cover, the untended seedlings that clutter backyards, roadsides and derelict agricultural land in America would be tolerated by few English landowners, private or public. (Lowenthal and Prince, 1965, p. 198)

Care for the environment is widespread in England and planning legislation gives the government wide powers. Permission is needed for any "development" such as the building of a garage for the family car(s) and any structural addition such as an extra room in the back garden or over the garage. There is also control of any change in the use of a building, such as the conversion of a dwelling house into flats or offices, even if no building work is necessary for such a change of use. Local governmental authorities have planning committees to which any applications for a change of use must be made and their activities are closely watched by residents, whose concern is in some places channeled by such organisations as "The Friends of Thistletown" and even on occasion by some public spirited citizens who form *ad hoc* groups to prevent some change in land use, such as the building of new houses or offices, which the local authority may favour. Not infrequently, such spontaneous expressions of local opinion are successful.

But it is not always so, for it may be that some old houses about to be demolished are beyond repair, especially if they were originally built without proper foundations. Historic premises may be officially named as "listed buildings" and in some towns there are also "conservation areas" within which no exterior change is permitted. Towns having extensive areas of eighteenth century and early nineteenth century houses, including London and Bath in England and Edinburgh in Scotland, retain their distinctive character, though former large town houses may now be offices, boarding houses or hotels, or even the headquarters of learned societies. At present there is an increasing interest in Victorian buildings and voluntary bodies such as the Victorian Society exist to advocate the preservation of Victorian work as already "historic" and even "antique." And people are willing to pay for their aspirations, as for example in the restoration of historic parish churches. In the town where the author resides the parish church raised £250 000 in four years for this purpose from its congregation and the generally friendly neighbours. Appeals for large sums are widespread and normally successful even in villages and small towns. For cathedrals, millions of pounds may be needed, as at Salisbury where an appeal was made for £6 000 000 to rebuild its famous spire. The heritage of the past is valued, not one hopes too complacently. Maybe some overseas visitors think that the English are making living museums of their towns, like old ladies who fill their sitting rooms with relics of a long past and supposedly happier time, but that is not the end of the planning story.

Local planning is not in general a major political issue. It may seem to be almost a hobby appropriate for the cultivated and aesthetically minded middle classes but it commands acceptance. Far more serious is the question of national planning and to many people major political issues are unemployment, housing and education. Modern technological advances have reduced the number of workers in industry, and hopes of former years that those deprived of industrial work could find employment in the service industries have not been realised. The sad plight of mining areas such as South Wales is well known but inevitable: now that the last mine in the

The passengers say: Sod the others
—let's keep going!
SOUTH
NORTH

Rhondda Valley has closed it is no comfort to the people to recall that in the 1850s people were provided with houses and other inducements to bring them into the valley. In the past, solutions lay in migration to such areas as London and the West Midlands where, William Hutton wrote in 1781, the metal trades of Birmingham "spring up with the expedition of a blade of grass, and like that, wither in the summer." True, but other blades replace them - or did in former times. Now even the West Midlands has a high rate of unemployment though for many years after 1945 it was a magnet for immigrants. And there are still more severe problems for the older coalfield and textile areas and for the northeast with its former dependence on shipbuilding and steel. Fifty years after the famous Jarrow march of the unemployed to London in 1936 a number of people marched again, saying that the problem of joblessness still remained. Under present circumstances is it really possible to believe that unemployment could be reduced from three million to one million within a few years, however vast any program of public works and housing might be? The problem of outworn industrial areas has long been known in Britain but the value of governmental action is a matter of political debate.

Mackinder's *Britain and the British Seas* (1902) has a plenitude of interesting ideas on many aspects of geography. Of special interest is his recognition of two major areas of England and Wales as they were at the beginning of the twentieth century (he deals separately with Scotland and Ireland). In Metropolitan England, broadly south of a line drawn from the Wash to the Severn estuary he recognises three main qualities: "firstly, nearly all the main roads and railways converge on London; secondly, the coastline, extended from Norfolk to Cornwall, everywhere looks across the Narrow Seas to the neighbouring continent; and thirdly, there are no considerable sources of motive power" (MacKinder, 1902, 232). Of these, the first remains true, the second is less significant now than in 1902 though the major airports are in the vicinity of London and the third is no longer of major significance in an age of universal electric power. But what follows is still relevant: "the commercial as opposed to the industrial control is there: and the whole has more or less of a residential character." Elsewhere he draws attention to the industrial strength of London, but that is only part of the later story for since his time there has been a massive increase of industrial activity within Metropolitan Britain. On Industrial England (with the whole of Wales), always aware of maritime contacts, he comments that there are "several important crossroads but a less immediate connection" with continental Europe. He goes on to say that Industrial England:

> has but a small proportion of the leisured classes, for rich and poor alike are workers, and as a result the prevalent opinions both in politics and religion differ not infrequently from those of the metropolis . . . the social life of Metropolitan England is old and aristocratic, whereas that of Industrial England is new and more democratic, for Industrial England, as a great community, is even more recent than the New England of America. (MacKinder, 1902, p. 202)

One might perhaps question the last statement and no doubt some of the very able economic historians who have worked in the universities of Mackinder's Industrial England would regard it as one of sweeping generalisations that (some) geographers are liable to make.

But how much of his general view is still valid? It is certainly true that within his Metropolitan England the political complexion is largely conservative and that in Industrial England radical views are more strongly represented. It is of course true that there is unemployment in Metropolitan England but on a smaller scale than in Industrial England. Average incomes are much higher in Metropolitan England but so too are prices, particularly for houses as noted earlier. Forecasting the result of the next general election has become a fascinating sport and a poll published in November 1986 gave the percentages here listed for political allegiance (Table 1).

The remaining voters (not included in the classes given in the Table) consist mostly of the 19 per cent who support the centre (Alliance) party, made up of Liberals, having a long tradition in British politics, and Social Democrats who have become a separate party only recently. In Scotland the political allegiance of those interviewed was 22 per cent Conservative and 49 per cent Labour, so it too could be regarded as part of the Labour heartland. Also of interest is that 15 per cent of those interviewed are supporters of the Scottish Nationalist party which favours more local autonomy. The comparable party in Wales had only 4 per cent of the voters. It must be stressed that these figures are just one reaction at one point in time in the autumn of 1986 and that the term "marginal constituency" is well known to all active party workers. When the results are declared during the long night following polling day in general elections, the trend of public opinion soon becomes clear, especially as some of the first results to be declared are from such politically marginal areas as Birmingham and the Midlands in general.

Probably Mackinder, a man devoted to politics and himself a Conservative member of parliament for a Glasgow constituency from

Table 1

Political Allegiances: 1986

	% Conservative	% Labour
A. The Tory Heartland (London and the south)		
London	43	37
Southeast	49	23
Southwest	48	27
East Anglia	48	30
B. The Labour Heartland (north of England and the whole of Wales)		
Northern	28	49
Yorkshire	31	50
Northwest (Lancashire and Cheshire)	35	45
Wales	32	47
C. The Marginal Midlands		
West Midland (including Birmingham)	44	37
East Midland (including Nottingham and Derby)	46	30

1910 to 1922, would have been fascinated by the "psephologists" who assess the ebb and flow of political opinion. His work as a politician has been closely discussed in Parker's *Mackinder: Geography as an Aid to Statecraft* (1982), and unquestionably Mackinder used his geography as a basis for political judgements. Regrettably, since his day there has been a wary approach by many geographers to broad national and international aspects of politics though at present there are many signs of renewed interest. Here, it is worth mentioning that at the meeting of the IGU Commission on Applied Geography held in 1972 at the University of Waterloo, problems characterising the relation between Canada and the United States attracted some interesting papers.

Geographers in Britain who were active in the first half of this century had a range of interests that would seem vast to their specialist successors of the contemporary period. It could be a useful exercise to look again at Mackinder's book of 1902 and list the aspects of the subject that are considered. Here attention will be given to just one of them, communications. In a career associated with three universities, Oxford, Reading, and London, he must have known the train service between Oxford and London expertly. Since his time the national system of motorways has developed, free of tolls and open to all who have passed a driving test. It is not a replacement for railways. Britain still has splendid "intercity" services, but the system has been pruned of many branch lines for economic reasons, though the argument for the retention for social reasons of some unremunerative lines continues. No industrial estate or other premises can have direct roadside access to a motorway because roads built during the interwar period from 1919-1939, such as the Great West Road out of London or the East Lancashire road between Manchester and Merseyside, proved to be magnetic in attracting new houses, factories and other industrial premises and direct access would have generated terrible congestion. The same feature was seen in the various "ring roads" beyond the built-up area of towns.

On October 29, 1986, the last section of the M25 motorway round London was opened with a severe speech by the Prime Minister on "carping critics" but this did not inhibit the *Guardian* newspaper from printing an article headed "Ring of No Confidence" on the following day. The M25 had taken 14 years to build and its total length is 120 miles. The *Guardian* writer, Terry Coleman, reported that he had driven round it in two hours and 20 minutes, during which he saw only five police cars but no example of dangerous driving. Conversely it took him 51 minutes to drive 14 miles through London suburbs to reach the motorway, constructed over much of its course as much as 20 miles from central London, which in his view has - with few exceptions - "the worst approach roads of any big city in the Western world." He goes on to say that "A man named Robert Moses built better roads for New York in the 1920s than we have today . . . You can get out of Manhattan to New Jersey or Long Island with reasonable ease. On highways running up the east and west sides of Manhattan you can get out fast into lower New York State and New England." Conversely the new English road has compensations for "how beautiful the countryside is, particularly to the south and west, in Kent and Surrey." But "it's absurdly far from the centre, which must be obvious even to those . . . who ensured by their protests that it should not be closer in." For all that, any land near the motorway on which new building, residential or other, is not inhibited by Green Belt legislation, has rocketed in value, with the figure of £80 000 an acre mentioned as a present price: little such land is available. This raises the problem long foreseen by those who have studied planning legislation: could the price for the preservation of England's green and pleasant land be stupendously high?

In the years that followed 1945 everyone knew that a return to the conditions of living known in 1939 was not possible. Now forty years later, it is clear that many mistakes have been made, that there are vast social problems in our cities, that it is as difficult as ever to make the right decisions in planning for the future. The hopes of some British geographers that they could be the great coordinators of the inevitable planning process have been modified by experience, for planning - as indeed many people realised at meetings of the IGU Commission on Applied Geography to which Peter Nash gave such good service - is a concern of politicians, economists, lawyers, architects, social workers and many others. Essentially it is a multidisciplinary activity, but if geographers retain their vital appreciation of man and the land, with landscape in town and country as a populated stage itself alive through its continual evolution, then they may do much to make a more humane world for future generations.

REFERENCES

Coleman, A. (1985) *Utopia on Trial: Vision and Reality in Planned Housing*, London: H. Shipman.

Coleman, A. and Power, A. (1984) "Trouble in Utopia," *Geographical Review*, 153: 351-62.

Freeman, T.W. (1962) Review of J. Gottmann, *Megalopolis*, *Geographical Journal*, 128: 534-35.

Gage, W.L. (1968) *The Comparative Geography of Palestine and Siniatic Peninsula*, (translation of part of *Erdkunde von Asien* by Carl Ritter (1866), New York: Greenwood Press.

Gottmann, J. (1961) *Megalopolis: The Urbanized Northeastern Seaboard of the United States*, New York: The Twentieth Century Fund.

Lowenthal, D. and Prince, H.C. (1965) "English Landscape Tastes," *Geographical Review*, 55: 186-222.

MacKinder, H.J. (1902) *Britain and the British Seas*, New York: D. Appleton and Co.

Mallory, W.H. (1926) *China: Land of Famine*, New York: The American Geographical Society.

Parker, W.H. (1982) *MacKinder: Geography as an Aid to Statecraft*, Oxford: Oxford University Press.

Ritter, C. (1822-1859) *Die Erdkunde*, 19 Vols., Berlin: G. Reimer.

Smith, G.A. (1909) *Historical Geography of the Holy Land*, 15th edition, London: Hodder and Stoughton.

CHAPTER 6

PERSPECTIVES ON APPLIED GEOGRAPHY IN THE CONTEXT OF INDIA

Mohamad Shafi
Aligarh Muslim University
Aligarh, India

INTRODUCTION

It was in 1960 that Dudley Stamp, the celebrated British geographer, published his book *Applied Geography* in the Pelican series, and since then the theme of applied geography has assumed a significant place in the geographical literature. At the 20th International Geographical Congress held in London in 1964, a new section was introduced on which papers were invited. This was the section on Applied Geography and I had the privilege of chairing this session. A good deal of enthusiasm and interest was shown by geographers in this section which finally culminated in the establishment of an IGU Commission on Applied Geography in 1964. Professor Phlipponneou (France), Professor Peter Nash (U.S.A.) and myself (India), among others, became members of this commission, and Professor Omer Tulip of Belgium was elected chairman.

The Commission held its first meeting at Prague in Czechoslovakia in 1965, and considered the origin and development of applied geography in socialist countries, in developing countries, and in countries with free enterprise economies. Also considered were such aspects as the interdisciplinary aims of applied geography and its relationship with public and private enterprise. The Czechoslovakian Academy of Science published the papers presented at the conference under the title "Applied Geography in the World" (Strida, 1966).

The second meeting of the commission was held in Kingston, Rhode Island, U.S.A., in 1966. This meeting was organised by Peter Nash. Every detail of the meeting, from the academic sessions to excursions, special dishes of lobster, to the cultural programs, was meticulously planned and supervised. It was indeed a remarkable meeting of the Applied Geography Commission. This Commission had thus begun its long and productive existence. It had one of the longest lives of any commission in the history of the IGU - three consecutive terms of four years each.

What is Applied Geography?

Peter Nash, convener of the 1966 Commission meeting, posed the question "What is Applied Geography?" to sixty-four American geographers prior to the Kingston meeting. Answers were communicated to Peter in writing. Almost all answers differed in phraseology, but they revolved round a single common theme. It can be summed up as "the application of geographical methods and techniques to the solution of practical problems." Geographers have well developed techniques of survey, analysis, interpretation and synthesis, and the application of these methods towards the understanding and interpretation of some of the problems of the world fall in the realm of applied geography.

Peter Nash perhaps posed this question in order to confront the widespread apprehension about the nomenclature of the term "Applied Geography." There was reluctance in some quarters to accept the term. It was argued that acceptance of this perspective would introduce a dichotomy in geography with reference to theoretical as opposed to applied geography. In pure science, the theoretical side is engaged in discovering laws while the applied side makes use of those laws for the benefit of mankind. In geography, where we take a holistic view of man and environment, no such dichotomy is permissible.

What is, perhaps, ignored in the above mentioned objection, is appreciation of the conceptual framework of applied geography. It is true that geographers attempt to understand and interpret the relationship between man and his environment in its entirety, both in space and time, but any spatial problem on the earth's surface, which is subjected to geographical analysis, mapping, interpretation and synthesis could fall in the realm of applied geography.

It should be realised that applied geography transcends the boundaries of the traditional individual branches of geography. The problem under study may relate to agriculture, industry, resources and their conservation, transport, recreation and leisure, population and environment, urban development or any other problem relevant to the society to which the geographer can apply his analytical techniques (e.g. statistical techniques and even computer programming) to make a diagnosis, or prognosis which may lead to prescription. Any or all of such research could fall in the realm of applied geography. Thus, the canvas of applied geography is vast, but the brush, which is used, is of the finest type so that features come out distinctly. Geographers should have the ability to produce realistic, dynamic and reasonably complete models of the system with which they deal.

The problems discussed at the second meeting of the commission at Kingston, Rhode Island, were selected carefully and the participation of specialists from related fields in their consideration showed that at least on some problems geographers must work as members of interdisciplinary teams.

APPLIED GEOGRAPHY IN INDIA

In the developing countries in general, and in India in particular, the most important problem is the increasing pressure of population on land resources. In many parts of the country, the population suffers from undernutrition and even malnutrition. Moreover, there is a growing demand on land for a variety of purposes which calls for a balanced use of land. Every hectare of agricultural land in the country has to be put to its maximum use. This obviously leads us to the problem of land use planning which involves, in the first instance, preparation of an inventory of existing land use based on a careful survey, and knowledge of the causes of those uses in the light of historical, physical, social and economic factors. Unless present use is known and understood, development schemes may cut across the basic structure of agriculture and society in a manner that may do more harm than good. Land use planning may not only result in a land classification based on fertility and productivity but also in a land capability classification.

Industrial expansion is making a heavy encroachment on agricultural lands in India. First class agricultural lands are being usurped by industrial establishments. Brick kilns are being set up on good quality land which result in derelict lands. The preservation of good quality lands for posterity is one of the country's most important problems. Applied geography has a distinct role to play in this field by application of the concept of "Potential Production Unit." There is need to develop land classification surveys on the basis of productivity of land at an existing level of technology. Such a survey may categorise land, say, in three types - A, B, and C. If it is supposed that the productivity of type B land is 1, that of A is 2 and of C as

0.5, and if it is further supposed that 1,000 hectares of land are needed for nonagricultural purposes, the utilisation of type A will actually deprive the nation of 2,000 hectares, while the use of type C only 500 hectares. Classification on this basis, will help to demarcate first class agricultural lands that should not be utilised for nonagricultural purposes.

Resource Management

India presently faces three outstanding problems in the management of resources: water management in relation to crops, utilisation of wastelands, and diversification of cropping patterns to increase production of oil seeds and pulses.

It should be noted that in spite of large investments made in the irrigation sector, and the phenomenal growth of irrigation in the country, during the past thirty years the return from such investments, in terms of output, is not satisfactory. Irrigated land should yield 4 to 5 tonnes of grain per hectare per year. At present, the yield is hardly 1.7 tonnes on average. In projects where appropriate water management and other cultural practices were maintained at optimum levels, yields have been satisfactory. Improvement in productivity per unit of water applied in areas already covered with irrigation arrangements should be put into effect. The immediate need is to reduce the gap between existing and potential irrigation areas in order to optimise agricultural production through better management of land and water. The objective of water management in irrigated agriculture is to provide suitable moisture and environment for the crops to obtain optimum yields with corresponding maximum economy. This is location specific technology and it is governed by the nature and extent of water availability, soil and climatic conditions, and the nature of the terrain to be irrigated. Thus, the situation differs from one area to another and herein lies the task of the applied geographer who should bring out clearly the areal differentiation, and offer proper solutions for efficient and effective management of land and water.

One of the important aspects where applied geography may play a positive role lies in the utilisation of wastelands in developing countries, including India. A land use survey based on remote sensing, aerial photographs and ground surveys along with other data, and its mapping and interpretation will go a long way toward putting land management on a scientific basis. Memories could be refreshed by recalling the monumental work done by geographers in Britain who conducted a land use survey of the whole country with the help of voluntary geography students from colleges and universities in the 1930s and during World War II when imports of foodstuffs were restricted. Britain was able to almost double her food production by scientific land use planning helped by these maps and their documentary interpretation.

Today in India, there are millions of hectares of land designated as "cultivable waste." Most of these lands suffer from excessive salinity and alkalinity. Alkali conditions are mainly due to sodium carbonate and bicarbonate, the exact quantity of which differs according to the nature of the soil, subsoil and level of the water table. Saline soils are characterised by the presence of sodium chloride and sulphate and have very little sodium carbonate and bicarbonate. They occupy a smaller areal extent than alkali soils. Sporadic work in the identification of saline and alkaline lands has been carried out on a collective basis in some parts of India and methods of reclamation have been devised. What needs to be emphasised however, is that the methods of reclamation are different for saline and alkaline lands. Even the National Commission on Indian Agriculture (1976) laments that no systematic attempt has yet been made to delineate these areas separately; instead they are mapped together. By detailed mapping, reclamation measures become more specific and easier to apply. Herein, lies an important task for applied geographers. At the village level, waste lands, in the first instance, could be identified on cadastral maps, available at a scale of 1:3,960. By precise soil analysis, alkaline and saline lands could be demarcated separately and then a cost-benefit analysis carried out to determine the most profitable use of such lands after reclamation - whether for crops, pastures or village forests to supply fuel, wood for implements and construction purposes. There is indeed a great potential in these lands, after reclamation, for augmenting the country's food production. Alice Coleman has rightly pointed out that the most vital need for the future is better analysis of land use surveys to give better insights into problems and opportunities as a basis for more effective and imaginative policies and action. This is a splendid opportunity for applied geographers to offer their skills and understanding for the common good.

Another important sphere where applied geographers can make a positive contribution in India is development of a basis for progress towards diversification of cropping patterns. The introduction of "New Agricultural Technology" commonly known as "Green Revolution" has made India self-sufficient in food grains. Not only has the import of food grains been stopped, the country is now in a position to export wheat and rice to other countries. The green revolution, while solving the food problem of the country has had its side effects. In the first place it has contributed to social tensions because the gap between rich and poor farmers has widened. Rich farmers can procure necessary inputs in adequate quantities at the right time, while the marginal farmers, unable to compete, have had to sell off their land and have fallen in the category of landless labourers. Moreover, the new technology has been most effective only with reference to wheat, so the production of pulses and oil seeds has not shown any positive change over the past thirty years. The static position of pulses becomes alarming when one considers it with reference to the population of the country which is overwhelmingly vegetarian and is deprived of adequate protein in its diet. There is an urgent need to study in depth existing cropping patterns in relation to physical, social and economic factors. A balanced cropping pattern needs to be evolved. Again, it should be determined to what extent the chemicalisation of agriculture in India is interfering with the environment, especially its long-term effect on soils.

Impact of Mechanisation on Labour

A relevant problem to which applied geographers in India have not paid attention is the spatial pattern of mechanisation in relation to absorption or displacement of labour. The country is now committed to the use of electronics in a big way, so much so that it is proposed to take it to the field and the farmer by the twenty-first century. A pertinent question which falls in the realm of applied geography is to investigate the relationship between mechanisation and rural employment. The common belief is that mechanisation leads to widespread unemployment. The problem to be studied in depth is whether mechanisation leads to rural unemployment. In the states of Punjab and Haryana, where new agriculture technology has been introduced on a large scale, labour is scarce in the busy agricultural season. For example, at harvest time it is available only at exorbitant rates and must be drawn from distant places.

Applied geographers need to investigate whether there is an inverse relationship between farm size and productivity, and whether this condition exists under equal fertility conditions. The Indian Planning Commission, based on data for 66 regions, has concluded that there is an inverse relationship between crop output level and employment. Is this because of mechanisation? What is the spatial relationship between growth and employment?

Distribution of Hunger and the Structure of Agriculture

Another problem that looms large on the Indian horizon is that while agricultural production is increasing, undernutrition and even signs of starvation in some parts of the country are apparent and the condition of the agricultural labourer, in general, has not improved. This situation calls for an in-depth study of the structure of agriculture.

Comparative studies of changing agrarian structure are important to understand the changing pattern of agricultural productivity as well as the impact of new development efforts. There is increasing population pressure on land resources in developing countries. One-third of the population in such countries is undernourished, and yet there is no overall shortage of food. There exists an imbalance in food consumption. For example, it is estimated that 60 per cent of the food produced in the world is consumed by the 30 per cent of mankind living in the rich industrialised world. It is not enough to increase food production in a country or in a region. Along with production, the problem of entitlement should also be solved. The major cause of undernutrition of a large section of humanity is that of unequal distribution of food and income. The problem of hunger lies not so much with the supply of food as with its distribution and availability to the poor. Production and consumption structures should be taken into account simultaneously. It is an important task for applied geographers to examine the entire food system in selected areas of the world, and to offer solutions to the food problem. In doing so, applied geographers will not only take into account problems of food production in relation to inputs, but also: (1) changes in land tenure policy and agrarian structure affecting food production, (2) socio-economic factors leading to changes in the pattern of food production and nutritional availability, (3) marketing, transport and storage in relation to food production, distribution and consumption, (4) food habits and new sources of food, and (5) government policies and practices affecting food production and distribution.

Another potential field for applied geographers lies in the realm of Dry Land Agriculture, particularly in India. About 70 per cent of the sown area in the country is rain fed, and this area produces about 42 per cent of the country's food grain - most of the coarse grains, pulses and oil seeds. The strategy for development of dry land agriculture is based on selection of micro-water sheds, water conservation and water harvesting technology. Application of geographical techniques could yield rich dividends. In the first place, the geographer can apply his technique of spatial analysis of productivity measurements; second, a detailed and precise inventory of land use at the micro-level should be carried out and interpreted. Third, the sites of farm ponds should be identified. Fourth, a detailed study of the infrastructure and supporting services should be made. Such a study would provide a powerful aid to scientific and other disciplines in comprehensive and scientific planning of dry land agriculture.

One of the basic principles of land use planning is the optimum utilisation of land in keeping with the needs of the physical, social, and cultural environment. To achieve optimum utilisation, a comprehensive land use survey at the micro-watershed level in the dry lands is needed. Such a survey should prepare a detailed inventory of existing use and nonuse of land, texture and structure of the soil, soil moisture, relief, storage of rainwater facilities, and the existing state of supporting services. On the basis of such studies a plan could be formulated, identifying land that should be devoted to crop production, pastures, social forestry and animal husbandry. Similarly, variations with regard to experience in tilling, sowing, water-harvesting, mulching and soil conservation need to be observed spatially, and at the micro-scale so that the rich indigenous experience in the form of different variables could be correlated with variations in levels of productivity, and a plan formulated to suggest changes for relatively low productivity areas to bring them up to the level of higher productivity areas. Again, loss of all food grains in India owing to faulty storage in dry land areas is estimated at 6.3 million tonnes a year, equivalent to about 100 million dollars. This is a substantial loss, but even this estimate is not based on any accurate survey. Geographers, by their techniques of spatial survey and analysis (statistical as well as computerised) could reveal the total loss accurately and also suggest methods for its minimisation.

Another problem of great significance in the developing world is that of raising agriculture productivity. Various techniques for the measurement of agricultural productivity have been developed. These techniques can be applied to delineate regions of high, medium and low productivity. There are various factors which affect agricultural productivity. Independent variables are inputs, like fertilisers, seeds, irrigation, facilities, agricultural implements, labour and credit facilities. At the micro-level, they could be examined in low productivity regions to determine variables that could be increased or even decreased in their application to increase agricultural productivity and thus minimise regional imbalances.

There is need for local surveys to determine farmers' perception of the changing environment and of their capacity to accept and adopt innovations in farm practice. In developing countries there is need to undertake studies of stability and change in cropping patterns, to identify cores and peripheries of important crops in order to assess the impact of price change on cropping patterns. Similarly, the location of agroindustries, based on resources of particular regions should be suggested so that the problem of landless labourers could be reduced and migration from rural to urban areas stemmed.

Population Pressure

One of the most serious contemporary problems facing the developing world is the increasing pressure of population on resources. Most population studies in the developing world fall in the realm of family planning. Perhaps this is because of the desire of planners and administrators to make a quick impact on the rate of population growth. The geographer looks at this problem in its totality and takes into consideration the quality, structure and distribution of population in relation to the socio-economic and political situation.

An important problem relating to this aspect is the quality of life, particularly in urban areas in the developing world. The number of slum dwellers has increased beyond any reasonable proportion. There has been tremendous migration from rural to urban areas in search of jobs. According to current estimates, at least one-fifth and possibly more than half of the 600 million urban dwellers in the developing countries live in slums or squatter settlements in cities. As time passes, the problem will become more serious. By the end of the century, the population of developing countries is expected to double and urban population may treble. Most urban dwellers, whether migrants or urban born, will be poor and living in slums. In 1973, about 27 per cent of Delhi's population of four million lived in unauthorised squatter settlements, compared to 11.4 per cent in 1958. Today, of the seven million inhabitants of Delhi, more than one-third live in slums and squatter settlements. Half of Bombay's population, of nine million, lives in structures which are without any infrastructural facilities and amenities. In Calcutta there are registered slums, for which the Calcutta Corporation collects taxes. There are also unregistered slums. About half of Calcutta's population lives in either registered or unregistered slums. The position in Madras is not much different. Broadly speaking, one-fifth of India's total urban population lives in slums, and, if one includes squatter settlements, the number would be much greater. Conditions in other developing countries are similar to those in India where there is an ever widening gap between demand for housing and its supply. It may however be pointed out that the information on slums and squatter settlements that is available is inadequate. Here lies a role for applied geographers in research on a very important problem facing the developing world in general and India in particular. The geographer, while examining the spatial distributions of slums and squatter settlements at the micro-level, may also investigate the specific causes of the push from rural agricultural areas, and suggest how local environments can be improved to accommodate the rural population with economic benefits. Moreover, it is important to determine what possible improvements can be carried out in the existing slums. There is an urgent need to collect factual data on the macro- as well as micro-level, and before the surveys are undertaken data collection procedures should be standardised to produce information of comparable quality and quantity. The collected data should relate particularly to density, age and sex structure, intra-state and inter-state rural and urban migration, health, nutrition, education, employment and income.

Rural Industrialisation

Another potential field for applied geography is rural industrialisation. Rural industrialisation can take place in two ways. First, through expansion of handicrafts, small scale industries and agrobased industries, and second, by promoting the decentralisation of industries from the urban to the rural sector. The location of such industries has to be well planned and include the development of infrastructure. To reduce the cost of the products of small scale agroindustries and to prevent piling of stocks, modern production techniques that improve the quality of goods and the productivity of labour need to be studied and applied to small units. A good link can be established by development of agroindustries along with agriculture, on the one hand, and by development of ancillary industries related to the production of spare parts for the organised large scale sector, on the other hand. In India, small scale units are located close to big cities, and the dispersal of these industries in the rural areas has not yet taken place. Technology in rural industries should be refurbished so that they can help meet export demands. Marketing channels need be revamped to protect artisans from exploitation. Development of industrial units in rural areas should be based on a spatial approach leading to a horizontal as well as vertical integration of units. The location of these units should be based on a careful survey of the resource endowment of each area. Infrastructure, in respect of accessibility, market, availability of raw material and the location of buffer-stocks of critical raw materials should also be thoroughly examined. It is thus obvious that building up a sound data base is necessary to facilitate proper policy formulation and evaluation. The Annual Survey of Industries in India covers only small scale units registered under the Factories Act. For these small scale units, the data given by the census are updated on the basis of only a two per cent sample. Thus, there is a need to establish a holistic picture of the area with reference to all types of rural industrial units. This may help identify channels for the flow of adequate data on a regular basis for policy formulation. The geographer by adopting the industry-cum-area-development-approach can take suitable projects and interface with planners, administrators and decision makers, and thus play a useful role in building the rural industrial base and generating employment.

Medical Geography

There is another field, which for want of a better term is designated as "Medical Geography," where great opportunities for advancement of knowledge exist for applied geographers. The medical profession is concerned with the treatment, eradication or control of disease, but the geographer examines disease in terms of the influence of specific environments on pathological phenomena, on the one hand, and the relationship between disease and the environment from one area to another, on the other. Thus a fruitful line of research is mapping the distribution of both disease and pathological factors at world, regional, and local scales. Physical human and biological factors play an important role in the occurrence of disease. Geographers may make sample studies to examine the extent to which a population is exposed to disease and analyse the influence of each environmental factor on pathological factors. It is common knowledge that diseases can be classified under two groups: communicable and noncommunicable. These, in turn, can be divided into two types: disease due to tissue changes which are reversible and those that are irreversible.

In all communicable diseases like typhoid fever, cholera, malaria and tuberculosis, the agent is introduced into the human body by inhalation, ingestion or through wounds or abrasions. Dr. Jacques May has done pioneering work in this field through his atlas of diseases. In the case of noncommunicable diseases, the physiological factors are important, such as the functioning of the liver under different climatic conditions, and the influence of heat, temperature and altitude on red blood cells. If such factors are analysed under different climates, the geographical pattern of degenerative disease such as goiter, arthritis and even cancer may become clearer. It would be worthwhile to mention the formation of a unit of geographical pathology by the National Cancer Institute in Bethesda, Maryland, where geographers work with epidemiologists, statisticians, nutritionists and endocrinologists.

There are numerous diseases related to dietary deficiencies of minerals and vitamins. Undernutrition and malnutrition lead to a variety of diseases. However, if food is derived from a variety of sources, caloric intake can be taken as a fair measure of the health of a population. In planning public health, surveys based on comprehensive questionnaires can yield data which can be mapped to show areas of various degrees of nutrition. Effective planning can thus be undertaken.

The influence of climate on health has long been an interesting area of study. Indoor climate has been modified by central heating whereby, from a single plant, a large building may be supplied with air heated by water, and cooled in summer with a controlled relative humidity. Dudley Stamp has raised the question, What is an ideal climate? Does it include a seasonal rhythm or a daily rhythm? This is a question that needs investigation. What is needed is to examine the effect of atmospheric conditions on the functioning of the human body. Thermal stresses may be assessed for given conditions of activity, clothing, temperature, humidity, air movement and radiation. Today, the geographical distribution of climatic elements that produce thermal stresses are not well known. At the same time it is recognised that human comfort and therefore stresses depend not on air temperature alone, but on the combination of temperature, humidity and air movement. The real task of applied geographers should be to look for an index of climatic stress and strain. Current indices are not so useful for the conjoint assessment of thermal factors in the environment. Lee has pointed out that the difficulty here lies not so much in calculating applied thermal stress as in determining the physiological significance of a given stress level and in deciding the relative contributions of different stress-producing components to that significance.

Recreational Geography

Another subject needing attention from applied geographers in India is recreational geography. Over the past fifteen years in developed countries, there has been a major expansion of interest among geographers in how people spend their leisure time. Adoption of a five-day work week, progressive industrialisation of India, increase in free time, and a relative increase in incomes, have all led to an increase in demand for opportunity to spend both short and long vacations away from home. The recreation aspect falls into several subsectors: health improvement, tourism and excursion. Obviously, this is an important sector of the national economy and it needs to be well planned.

The recreational system is dependent on common natural resources, and at the same time it is independent with its own system of institutions, administrative bodies, planning systems and material supplies. The recreational system therefore should be analysed in the light of available resources and their share in the present and future development of the system. It should be mentioned, however, that demand for recreation is generated differently in different places depending upon the degree of concentration of population and the availability of free time. The climatic factor in India is an important element in planning recreation. In May and June, the north Indian Plains become very hot and there is an exodus of people to hill stations in the north or toward places in south India. Indians often use their leisure time to make pilgrimages to holy places, or to visit such places at the time annual fairs are held. Many of the natural, cultural and historical sites are quite unique and can be used for the organisation of recreational activities. Recreational resources should be considered as a group of sites offering cultural, architectural and historical experiences, and they should be considered as a recreational resource only if they satisfy the recreational demands of a large number of people over a specified period of time. These recreational resources can be grouped under three headings: (1) intensively used resources ready for development owing to recreational assets, (2) extensively used resources, and (3) resources not used.

It is suggested that the major function of geographers in the development of recreational geography is to delimit recreational regions on the basis of actual and potential use. The spatial distribution of recreational activities must be carefully examined and may change in the course of time. With improved data and with a more effective use of questionnaires and interview techniques, it should also be possible to estimate the monetary value of recreation.

Tourism and recreation at both the domestic and international level have grown rapidly in India. At the international level there has been a substantial increase during the last two decades. The number of foreign tourists in India has increased from 17,000 in 1951 to 7,650,000 in 1979. To promote tourism and recreation facilities, the industry's infrastructure has to be carefully examined and planned. The spatial distribution of hotel accommodation, internal transport, particularly air services, comfortable surface transportation and airport facilities, have to be studied, mapped and deficiencies pinpointed. Development of recreation facilities must not adversely affect either the environment of the surrounding or local cultures or ethos of the places of tourism. This requires a careful inventory and mapping of areas where tourism is to be promoted. India, with its long, rich cultural heritage and variety of landscapes, provides a favourable ground for development of tourism. Ancient historical monuments and museums, places of pilgrimage of different religions, health resorts and hill stations, and old and new cities with a variety of urban landscape provide a sound basis for the attraction of tourists. Once the importance of tourism is fully grasped, it should constitute a major theme in applied geography in India. Its development will promote national integration and international understanding, provide new employment opportunities, remove regional imbalances, and open up new growth centres in the interior of the country.

Transportation Geography

Another subject needing attention by applied geographers is transportation geography. Geographers have to look at transport problems in all their aspects, particularly freight and passenger traffic, industrial and agricultural needs, rural and urban requirements. The problem of urban transport is becoming more acute day by day owing to a great increase in need, on the one hand, and to a shortage of resources, on the other. Several factors have in recent years made the situation critical, namely (1) the growth in population, and (2) migration from rural to urban and metropolitan areas in search of jobs. As a result, the central transport system in big Indian cities is bursting at the seams. Applied transport geographers should examine the entire transport system at the macro- and micro-level to remove bottlenecks hindering movement of industrial and agricultural goods. By projecting transportation requirements into the future, they may be able to see how additional capacity can be created in transport to meet the requirements of anticipated traffic. In the Indian context it is important to give special attention to the transport needs of remote and isolated areas.

FUTURE RESEARCH

An important task for geography is the initiation of academic programs in futurology/futures research. In fact, many short- and long-range studies have been commissioned in India, and eighteen such studies have been completed and printed. Some of the topics studied relate to an outlook for India's future for (1) housing, (2) food production, including rainfall uncertainties, (3) urbanisation, including new growth centres, and (4) transportation. Research in futurology, in fact, has to be tackled not only on the macro-level but also at the micro-level, and what is more important, is its consideration spatially, at the block and village levels. It must take into account the elimination of regional disparities and the fact that existing as well as future growth must be accompanied by social justice. An inventory of existing facilities should be prepared and planned at the micro-level and the projected requirements should be analysed and planned. Accomplishment of these tasks will require the services of geographers in a big way. Applied geography will yield rich dividends if it focuses attention on ocean futures, minerals, energy futures, steel, paper and glass futures, sugar, drug and textile futures. But there are certain important items which brook no delay in the Indian context. The spatial availablity of clean drinking water in urban as well as rural areas, availability of adequate nutrition to individuals, proper housing facilities, provision of recreation facilities with associated improvement in the quality of life in urban areas in the face of increasing population, are just a few of the problems in India to whose solutions applied geographers can make an important contribution.

It has been shown in this essay that applied geography stands in the service of society - in the fields of agriculture, industry, transport, recreation, disease, and in the location of settlements and trade centres, as well as in many other fields. What will be the future of applied geography? One could echo the statement of James and Jones (1957), namely, that applied geography as a distinctive field will grow or decline in proportion to the contribution it makes in understanding society's problems and in finding solutions for them.

REFERENCES

Brown, E.H. (ed.) (1980) *Geography, Yesterday and Tomorrow*, New York: Oxford University Press. See chapters on: "Land use survey today and tomorrow," by Coleman, A., "The geography of leisure and recreation," by Coppock, J.T., "Medical Geography," by Howe, G.M.

Brown, L.R. and Eckholm, E.P. (1974) *By Bread Alone*, Washington, D.C.: Overseas Development Council.

Biswas, M.R. and Biswas, A.K. (1979) *Food, Climate and Man*, New York: John Wiley and Sons.

Christians, C. (ed.) (1967) *Colloque International de Géographie Appliquée*: 3rd Réunion de la Commission de Géographie Appliquée, Liège: L'université de Liège.

Gerasmimov, I.P. (1983) *Geography and Ecology*, Moscow.

Griffiths, J.F. (1966) *Applied Climatology*, London: Oxford University Press.

Institute of Economic Growth (1982) *Relevance in Social Science Research, A Colloquium*, New Delhi.

James, E. and Jones, C.F. (eds.) (1957) *American Geography Inventory and Prospect*, See chapters on "Physiological climatology," Lee, D.H.K., "Medical Geography," Jacques, M.M., Syracuse: Syracuse University Press.

Kahn, H., Brown, W. and Martel, L. (1976) *The Next 200 Years*, New York: Morrow.

Lenihan, J. and Fletcher, W.W. (1975) *Food, Agriculture and the Environment*, Vol. 2, London: Blackie.

Michel, A.A. (ed.) (1967) *Proceedings of the Second International Meeting: Commission on Applied Geography of the IGU*, Kingston: The University of Rhode Island.

Preston, R.E. (ed.) (1973) *Applied Geography and the Human Environment*: Proceedings of the Fifth International Meeting, Commission on Applied Geography, IGU, Waterloo: University of Waterloo, Department of Geography, Publication Series 2.

Progress Publishers (1982) *Recreational Geography of the USSR*, Moscow.

Progress Publishers (1983) *Society and the Environment*, Moscow.

Stamp, L.D. (1960) *Applied Geography*, Harmondsworth, Middlesex: Penguin.

Stamp, L.D. (1962) *Land of Britain, Its Use and Misuse*, London: Longman.

Stamp, L.D. (1963) *Our Developing World*, (2nd. ed.), London: Faber and Faber.

Stamp, L.D. (1964) *The Geography of Life and Death*, Ithaca: Cornell University Press.

Stamp, L.D. (1964) *Some Aspects of Medical Geography*, London: Oxford University Press.

Stamp,L.D. (1969) *Land for Tomorrow*, Bloomington, Indiana: Indiana University Press.

Strida, M. (ed.) (1966) *Applied Geography in the World*, Proceedings of Prague Meeting of the IGU Commission on Applied Geography, Prague: Czechoslovakian National Academy of Science.

Swaminathan, M.S. (1982) *Science and Integrated Rural Development*, New Delhi.

CHAPTER 7

THE NEED FOR RESEARCH ON THE CONTRIBUTION OF MIGRANTS TO HOST POPULATIONS

Wolf Tietze
Editor, *Geojournal*
Helmstedt, West Germany

INTRODUCTION

Over the centuries the majority of immigrants to North America were farmers, sailors and soldiers - young people of rural background with little or no vocational training. In contrast, the twentieth century immigrants - still by far mostly from Europe - were characterised by a large proportion of highly qualified urbanites, including many top-ranking scientists and artists. Little attention has been paid so far to the benefit North America enjoys from this recent migration, which certainly is not properly comprehensible by simply applying and evaluating the usual demographic statistics.

Such is the case, because demography, including all its subdisciplines like population geography, population statistics and their interlinkages with both urban and regional planning, is always busy "counting sheep," and has developed many tools and methods for this purpose. Applying these constantly improved instruments permits rapid and even more accurate prediction of population growth and decline, on the one hand, and effective utilisation of ever smaller samples, on the other. Computers have contributed substantially to a high tide of progress in this field. The general enthusiasm for this welcome development is quite understandable. Science, however, cannot flourish on emotional fire (enthusiasm) alone. A closer look at the field of demography, however, quickly reveals that all of this progress is restricted to *quantitative aspects* only, while *quality* remains the Cinderella in the family of the population sciences - much to the distress of planners, historians, sociologists, and geographers who focus attention on the particular identity of distinct regions.

Politicians and businessmen, too, might be better served if a fuller understanding of population structure and dynamics was at hand. Admittedly, there are usually data available on sex ratio, age structure, and even on educational background and vocational training. Still, the extent to which qualitative data exist in sufficient detail usually appears to be on a rather archaic level. The politically fashionable ideology of protecting the private sphere more than is necessary impedes collection of relevant data. It is surprising how infrequently the popular slogan "Quality of Life" triggers serious contemplation of the "Quality of Humanity" and on its underlying crucial interdependencies. Isn't "Quality of Humanity" an essential component of "Quality of Life?" The alternative strong Marxist attitude in the theory of science has certainly not been a fertile field for methodological refinement for research on human quality. The elite/proletariat polarity so far more a matter of emotion and ideology than of ratio.

COMMENTARY

Obtuse "sheep counting" plays a particularly notorious role in the development of international migration laws and statistics. The generally acknowledged principle of "Legal Equality" represents a major achievement in the century-old struggle for increased implementation of human rights because it stands in opposition to any selective treatment of people. On this pretext, virtually all world governments were remarkably obstinate when in the 1930s and early 1940s the Jewish population of Europe, in a dramatic state of emergency, tried to escape the holocaust. Unexpectedly, possibly - undesirably in any case - the existence of a ubiquitous antisemitic attitude became evident, although hardly anywhere admitted, of course. This fact not only contributed to an increase in the number of desperate victims, but it served as an excuse to the perpetrators. Imagine this perversion!

The shock of the holocaust experience to those concerned and to those just rescued, as well as to those involved, innocently, yet burdened by collectively being related to the murderers, has been enormous. This shock is not digested sufficiently to allow unemotional investigation of this extraordinary historical event in a manner facilitating a search for lasting conclusions. In view of this task, which may be accomplished when the time is mature, a few thoughts are offered here for further contemplation.

The reproach to demography for placing too much emphasis on quantity and neglecting the quality of humanity demands that the field develop new research goals and methods. For example, any individual, as a component of society, either subtracts from or contributes to the society to a constantly varying degree. This input-output balance may serve as an arbitrary indicator of human quality. It is easy to understand that any individual needs many years to mature and to learn. During the course of this process, the input by society is rapidly evolving to a turning point where the efficiency of the respective individual starts to replenish the batteries of society. This efficiency will normally pass through a more or less distinctive period of climax, and eventually decline and drop below the level of balance again. Not much imagination is needed to understand that an infinite variety of such input-output diagrams is possible.

When this perspective is applied to the European-Jewish refugees/immigrants to North America in the 1930s and 1940s, "efficiency-curves" are to be expected with quite impressive above-balance sections which could hardly have been produced to this degree by any other group of migrants in human history. The reasons are:

1. The majority of these migrants represented an optimal age group, in terms of economic efficiency.
2. Their level and variety of skills were far above any average. Almost all of these migrants were highly motivated intellectuals, flexible and adaptable urbanites, multilingual, and in command of an internationally productive network of relations and experience.
3. Besides businessmen and lawyers there were thousands of highly reputed scientists, musicians, writers and poets, including many Nobel laureates and globally leading representatives of the fine arts.
4. Virtually all of these migrants were extraordinarily industrious, talented and determined.

Isn't it true that the holocaust tragedy overshadows the brilliance of this phenomenon? It could be asked whether most contemporary governments were concerned mainly with a fear of competition instead of with the human benefits of accepting the outstanding challenge? Indeed, the North American countries, Canada, and much more so, the United States of America, have more than any other countries of the world harvested this outstanding brain drain. The United States has benefited the most, and has gained an immeasurable yield from immigrants. The dimension and impact of the process just described may to some extent be surmised by observation of its silent fading in recent years, a fading caused by the natural aging and passing away of the injectors. The contours of the phenomenon are, of course, becoming obscured now by the takeover by subsequent generations.

There is another case that has been missed by a quantity-minded population science and its immigration statistics. It is similar in dimension as far as consequences are concerned. Numerically, however, the group of immigrants referred to is much smaller, numbering only about some 200. They arrived in the USA shortly after World War II. One can almost say that they went voluntarily, especially when compared with that of their former colleagues (almost the same in number) who were forced to work in the Soviet Union. These people represent another unusual example of brain drain, and one with epochal consequences. They laid the foundations of the jet, the rocket, and the space age, both in the West and the East. To quantifiers this small group of people may represent a negligible order of magnitude. However, the elite character of these few scientists and technicians, combined with the working conditions provided by the respective host governments, truly initiated mankind's march to new horizons.

Both examples given here exemplify virtually unplowed fields for research, although they obviously contain a wealth of intriguing detail in their own right. More thorough understanding of the qualitative dimensions of migration should be an objective of future research projects. Enlightening results can be predicted with certainty.

PART III

GEOGRAPHY AND HUMAN VALUES

CHAPTER 8

REGIONAL GEOGRAPHY AND PETER NASH

Aubrey Diem
Department of Geography
University of Waterloo
Waterloo, Ontario

THE VERCORS MASSIF: A GEOGRAPHICAL ANALYSIS OF A FRENCH ALPINE REGION

Not long ago, my wife Heather and I were bicycling south of Grenoble on our way to the delta of the Rhône and the Languedoc coast. Following the Isère Valley, we skirted a pre-Alpine massif of Jurassic limestone and stopped to camp at Pont-en-Royans. Having some time to spare before nightfall, we unloaded the bikes, set up the tent, and then headed a few kilometres up the Bourne valley where the river had cut a deep canyon through the layered white rocks of the Vercors.

I then realised that Peter Nash had travelled through and written about this area just after World War II when he was a soldier in the U.S. Army. In fact, I had read his Master's thesis, *The Vercors Massif: A Geographical Analysis of a French Alpine Region* (1946). When I returned to Waterloo in the autumn, I asked Peter to give me his copy once again so that I could reread it. This time I studied it with special interest as I had now been through a part of the area that he had discussed. In many ways *The Vercors Massif* was similar to, though far more detailed than, my own Masters thesis on another Alpine region, the Valle d'Aosta in northwest Italy.

I can best describe *The Vercors Massif* as an example of classical regional geographical writing. Peter examined and analysed the area's physical, historical, cultural, and economic geography. As well, there was an important section describing the terrible damage wrought to the Vercors by the German Army during 1944. After finishing the thesis and learning much that I was not aware of when I made my brief stop at the edge of the massif, I realised that I had read the only work in English, up to that time (1946), that concentrated on the Vercors. It was an unknown area for most geographers who lived outside Europe. Only fifteen of 121 works cited in Peter's bibliography were in English and they were of a general nature. The majority of the other entries were in French; however, because of Peter's schooling in Germany, France, and England, he was fluent in German, French and English and thus had access to sources in those languages.

At a young age, by doing field work and research for his Master's thesis, Peter Nash laid an important part of the foundation for a lengthy and illustrious career, one that ranged over much of the subject matter of the field of Geography. His eclectic approach to the subject was rooted in a solid foundation in the basic elements of the field. What a contrast between Peter and so many of today's younger geographers who are so specialised that they have little idea of the fundamental physical and human knowledge that provides the groundwork for geographical inquiry. Few are fluent in any but their own language. Most know little about areas of the world other than that in which they have been brought up. They have been "trained" to be geographers; however, they have not had a geographical education.

A GEOGRAPHICAL EDUCATION

An understanding of many subjects is vital for a geographical education. I would like to discuss three that for the most part have only been cursorily studied. They are: (1) an understanding of human perception in regards to place; (2) knowledge of the psychological nature of humans as a factor in explaining behaviour; and (3) the ability to integrate disparate subjects so that cause and effect may be considered within a regional setting. My examples are personal and are meant to stimulate the reader to search for related situations pertinent to him or her.

Perception

Firstly, I will look at perception. Having grown up in Detroit, I was extremely interested in Bunge's analysis of the Detroit milieu in his study of the Fitzgerald Community. Among other things, he was critical of the playground facilities in the elementary schools that were in black neighbourhoods. According to him, they were substandard and unfit for recreational needs. A caption from one of the photos of a playground stated:

> The geographic concept called "land use" is normally restricted to property mapping. But land is also used directly by humans. What is it that the human child in Fitzgerald actually touches? Is this a suitable surface for human contact, or is it just cheap, easy to maintain, easy to drain? Or is it deliberately inhumane so as to discourage after-school use? Would anyone want to picnic here? (Bunge, 1971)

To be sure these playgrounds are forty-five years older than when I used them, and have obviously changed. Nevertheless, Bunge's perception was totally different from my recollections. I had grown up on these rough fenced-in fields, playing baseball, touch football, soccer, and numerous invented games, not only during the school term but also during the summers of my youth. These modest playgrounds were open spaces that enabled my friends and I to experience a great *joie de vivre*. They were an essential part of our maturing into responsible adolescents.

Geographers, because of their particular education, perceive space in a certain manner. And yet perceptions other than geographical are as valid and can be intensely meaningful. For example, artists, musicians, writers, and poets portray space from a point of view reflecting their interests, rather than from a preconceived set of criteria. A number of illustrations are evident. One such expression that has always struck me as extremely beautiful is Don Maclean's song *Starry Starry Night*. It immediately floods my mind with the tormented life that Vincent van Gogh must have experienced while painting, especially during his stay at Arles in Provence, southern France. Ann Mortifee sings it and interprets the lyrics most meaningfully:

> starry starry night
> paint your palette blue and grey
> look out on a summer's day
> with eyes that know the darkness in my soul
> shadows on the hills
> sketch the trees and the daffodils
> catch the breeze and the winter chills
> with colours on the snowy linen land
> now I understand what you tried to say to me
> how you suffered for your sanity

how you tried to set them free
they would not listen
they did not know how
perhaps they'll listen now

starry starry night
flaming flowers that brightly blaze
swirling clouds in violet haze
reflecting Vincent's eyes of China blue
colours changing hue
morning fields of amber grain
weathered faces lined with pain
are soothed beneath the artists's loving hand
now I understand
what you tried to say to me
how you suffered for your sanity
how you tried to set them free
they would not listen
they did not know how
perhaps they'll listen now
for they could not love you
but still your love was true
and when no hope was left inside
on that starry starry night
you took your life as lovers sometimes do
but I could have told you Vincent
this world was never meant
for one as beautiful as you

starry starry night
portraits hung in empty halls
nameless heads on frameless walls
with eyes that see the world and can't forget
like strangers that you've met
ragged men in ragged clothes
a silver thorn a bloody rose
lies crushed and burning on the virgin soil
now I think I know what you tried to say to me
how you suffered for your sanity
how you tried to set them free
they would not listen
they're not listening still
perhaps they never will

Though not accepted at first, and totally at odds with a formal geographic perception, Van Gogh's impressions of the landscape, vegetation and people of Provence emerge when one thinks about southern France. In fact, for many, the paintings that he completed while at Arles have become the visual embodiment for this region of the world.

A more modern perception of place may be experienced through the songs of Bruce Cockburn. He has written and sung about many countries, including Canada, Italy, Japan, Nicaragua, and Germany. One of his most recent efforts is *Berlin Tonight*, a song recording his wintry impressions of this divided city:

> dull twilight spits hesitant sulphur rain
> sky been down around our ears for weeks
> only once - gap glimpsed moon over that anal-retentive border wall
> as we laughed through some midnight checkpoint under yellow urban cloud
>
> weeks of frantic motion - petrol veins of europe pumping
> through scratchy acid-bitten transparent winter trees
> through brownish haze that makes a ghost of the horizon
> i'm rushing after some ever-receding destination
>
> berlin tonight
> table-dancing in black tights
> waving a silver crutch in the blue lights
> shape changing over glass
> on the front line of the last gasp
>
> green shoots of winter wheat and patches of snow
> russian walks dog in saxon field
> from the top of a solitary tree like the one on the flag of lebanon
> unblinking eye of hawk follows traffic on the autobahn
>
> tank convoy winds down smokestack valley
> proud chemical pennants wave against the sky
> turret gunner laughs when I throw up my hands
> I'm all glasses and grin to him under my commie fur hat
>
> berlin tonight . . .

Psychology

Another significant element that geographers have neglected is psychology. Not only must we know about the psychology of Western Europeans and North Americans if we want to shed light on why the earth is used and abused the way it is, we must also understand the arcane psychological impulses of, among others, Moslems, Jews, Asians, Africans, and Russians. Will we ever truly comprehend the inner forces that drive human activity? Regardless, we certainly have to know more than we do at present if we are to grapple with the multitude of problems that are engulfing the planet.

Why, for example, have humans, throughout history, left behind or refused to address the immediate problems that they have created and instead searched for new horizons to conquer? Certainly, solving or ameliorating the social instabilities of the United States is far more significant to the survival of Western Civilisation than shooting objects into space.

Whereas humans could strive in a positive constructive collective manner towards a better life for all earth dwellers, history has shown that a negative, destructive selfish approach has predominated. Not only is the past studded with murderous wars; it is also characterised in peaceful times by individuals who strive for personal gain regardless of the oppressive consequences to others.

One particularly odious example that has affected the economies of many nations is the illegal drug industry, a disparate, international consortium that has profited to the tune of hundreds of billions of dollars annually, and threatens the very existence of Colombia as a sovereign state. Another example, is the "legal" sale of arms by individuals and nations to anyone or any country that can pay for them. The result is bizarre to say the least. For example, on one hand, the World Bank provides millions of dollars for economic development in India. On the other hand India uses billions of its own money to purchase Soviet MiGs and tanks, English Jaguars, French Mirages, German submarines and Swedish trucks and howitzers.

Bertold Brecht in *The Threepenny Opera* has tried to alert us to the psychological inconsistencies of the human condition; however, even reviewers of his work fail to grasp the message. In a recent commentary of the Berliner Ensemble's production of this famous opera in Toronto, the *Globe and Mail*'s reviewer discussed the influences on Brecht's life, and wrote:

> But it is also true that that time is long past and the desperate world that created this Macheath and Peachem and their nimbus of exploited whores and panhandlers is gone as well.

If only this were so. In contemporary societies there are no shortages of Threepenny Opera characters. Their roles are played by such persons as corrupt government functionaries, false academics, Mafia types, zealous religious leaders, pornographers, murderous generals, unscrupulous executives, and those parasites (the fake beggars in Threepenny) who for no legitimate reason, take from, but give nothing back, to humanity.

On a more personal level, the weaknesses that ooze from Brecht's players are found in all of us. The characters that he brings on stage have been around since humans have inhabited the earth, and they will remain as long as civilisation endures. By exposing the hypocrisy that permeates mankind Brecht forces us to probe the deepest recesses of our minds and to continually evaluate the religions, political philosophies, ideas, "truths," and psychological defences that have enabled us to survive by sublimating reality.

Integration of Subjects

For the geographer, the challenge of understanding human perception and the psychological nature of humans is only the first step. Both, to be meaningful, must be combined with other subjects and analysed within the region. Formerly, this meant looking at limiting physical or cultural boundaries. Today, because of the integration of economies and a broader diffusion of culture, we must look at individual regions within the broader outline of "The World Region" to have meaningful results.

Geographers, in order to retain their unique point of view, must integrate physical and human factors to provide a useful and interpretative source of information, whether for government, business, or students. The region can be activated only when the relevant systematic elements have been plugged into it. By themselves, the systematic studies are sterile and inert, but once placed in the matrix of the region, they combine and interact to animate and expose its unique characteristics.

Classic geographical education is unique in combining the study of physical and human elements of the earth's surface. By overlaying these varied elements, including perception and psychology, a multidimensional "cartographic" type of interpretation emerges. The integration of systematic geographic subjects is the catalyst that results in a geographic viewpoint. In essence, this is what regional geography and Peter Nash's Master's thesis are all about.

I would be naîve to believe that all that is necessary for a peaceful and fruitful life for the billions of earth dwellers is a classical integrative and holistic geographic education. But I am not naîve in stating that without such an education, leaders and policy makers of the world's nations are ill-equipped to anticipate and react intelligently to the multitude of economic, political, and environmental problems that threaten to engulf human activity.

REFERENCES

Bunge, W. (1971) *Fitzgerald: Geography of a Revolution*, Cambridge, Mass.: Schenkman.

Nash, P.H. (1946) *The Vercors Massif: A Geographical Analysis of a French Alpine Region*, Los Angeles, California: University of California at Los Angeles Archives, unpublished M.A. thesis.

CHAPTER 9

SOCIAL POLICY, IDEOLOGY AND HUMAN SETTLEMENT PRACTICE

Demetrius Iatridis
Social Planning
Boston College
Boston, Massachusetts

This paper stems mainly from recent field trips with my students to Cuba and the People's Republic of China. We have completed field studies in Cuba every year for the last nine years and in the People's Republic of China each of the last two years, as part of a "Comparative Social Planning" seminar. In each field trip to these, and other, countries we visited different human settlements (planned, unplanned, urban, rural, small, large, as well as neighbourhood, metropolitan and megalopolitan areas), talked about "Ekistic" policy and practice with our professional counterparts, visited planning agencies, housing projects and community centres, talked with families who own and rent their dwellings, and observed actual ekistic conditions and facilities. Moreover, we tried to understand the role and importance of human settlements in over-arching national development efforts and priorities.

Juxtaposing ekistic experience in the U.S.A., and other market nations, with human settlement policies in Cuba, China, and other nonmarket nations, proved to be an insightful undertaking that sharpened and increased our understanding of ekistic theory and practice. This presentation centres mainly around understanding which resulted from comparing human settlement practice in the two systems.

IDEOLOGY AND HUMAN SETTLEMENTS

The most fundamental realisation that emerges from this comparative analysis is that the form and structure of human settlements reflect the larger ideological, socioeconomic and public administration fabric of the nation - particularly its mode of socioeconomic relations and societal arrangements (Figure 1). This may seem to be a fairly innocuous statement, but its theoretical and practical ramifications and implications are far-reaching. The notion, for example, that human settlements and the organisation of space reflect the ideology and the socioeconomic mode in which they are found suggests that their form and structure is not an inevitable consequence of unavoidable technological growth forces. Rather the form and structure of human settlements reflect accurately the ideological and socioeconomic relations dominant in a society. Instead of viewing human settlement as an inevitable consequence of industrialisation or other technocratic assumptions, the approach which views human settlements as part and parcel of the socioeconomic fabric of society highlights the influence of prevailing values, priorities and social policy priorities in a given country. By focusing on the broader ideological and social policy priorities one perceives the structure and function of human settlements as man-made and, therefore, subject to improvement through ekistic policies.

This approach underscores also the differences between neighbourhoods, between city and suburbs, between urban and rural areas, between regions and between countries. Moreover, this type of analysis unravels the causes of uneven ekistic development and inequitable distribution within cities and between human settlements. For example, differences and inequities in human settlement and between urban-rural areas seem to grow continuously in the U.S.A. and other market nations. In stark contrast, human settlements in nonmarket nations, like Cuba and China, tend to develop more evenly and equitably, even reducing distributive inequities between urban and rural areas through ekistic practice.

Ideologies are perceived in this paper as social processes, not as ideas (or political thoughts) possessed. Public or individual action is almost impossible without ideological underpinnings; hence, ideology is, in a practical sense, the prevailing public policy. Facts and action are meaningful only in the context of an ideological framework.

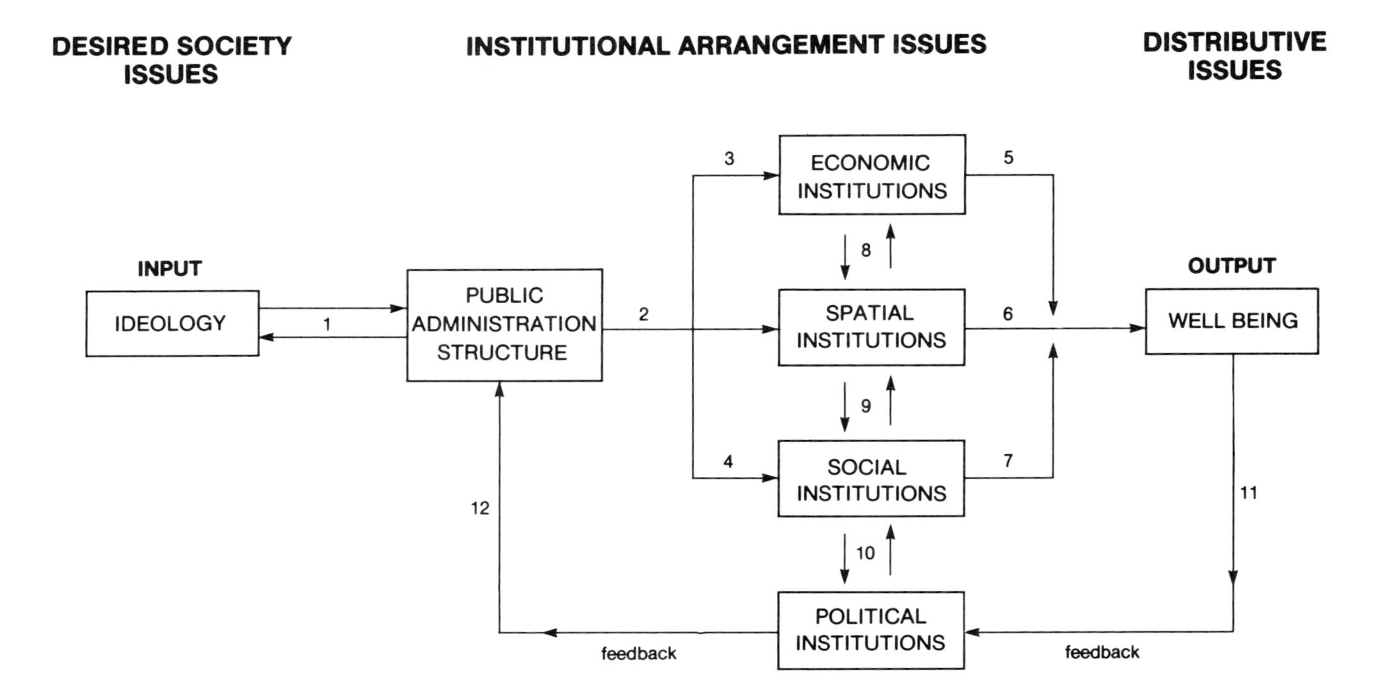

Figure 1: Ideology in Ekistic Theory and Practice

FACTS, VALUES AND ACTION: A BEGINNING MODEL

Another fundamental realisation from this comparative analysis is the dominant role of values in shaping human settlement policy and practice. Ekistic practitioners recognise by-and-large that facts, values and action constitute three major components of the ekistic process. In practice, however, we tend to be more aware of our emphasis on facts and action and less conscious of the influence of our own values and those of our clients upon the choice of options and priorities. Yet we recognise that facts are really meaningful only in the context of values, in the sense that, for example, six per cent unemployment or .55 Gini coefficient of income distribution become critical or preferable only in the context of prevailing values. Moreover, action can be legitimised or justified only in the context of values, as for example, an ekistic policy recommendation to achieve a balanced, mixed neighbourhood or town. Hence, all three major components of the ekistic process usually involve values.

This is also true of the major questions practitioners face in ekistic practice (Figure 1): What is the desired, ideal human settlement (I, Figure 1)? What spatial and institutional arrangements are necessary to produce it (II, Figure 1)? And, who gets what in terms of housing, public amenities, facilities, equipment and space (III, Figure 1)? Answers to the first question can be conceived as the "input" of the model in Figure 1, answers to the second question influence the "institutional arrangements" part of the model, and answers to the third question about distribution affect the "output" or "well being" conditions in a human settlement (Figure 1).

There is a cause and effect relation between ideology and the kind of public administrative structure in a given society (Link No. 1, Figure 1); between public policy and social, economic and spatial institutions (Links 2, 3, 4, Figure 1); between these institutions and well being conditions (Link 6, Figure 1); and between actual well being conditions and the political institutions designed to maintain equilibrium and resolve conflict (Links 11, 12, Figure 1). These relations are further illustrated in the sections which follow.

Ekistics and Central Command Planning

The third realisation concerns the distinctly different role of ekistics in the development effort of market and nonmarket countries. In less developing, nonmarket nations, ekistics becomes a part and parcel of central, over-arching, national development planning. By contrast, in advanced market countries, human settlement policy plays only a peripheral role in the development effort, which is mainly a function of free market forces at the local level rather than of national planning.

Both in Cuba and China ekistic policy is integrated, for example, with full employment planning, economic production, and equitable social development in the context of a central national development plan. Moreover, human settlement policy has become an effective tool for balancing and integrating rural-urban differences mainly through creation of new communities. Communes in the People's Republic of China, which as new human settlements became the basic social and production unit in rural areas, and "communidades de nueva formacion" (newly formed communities) in Cuba, which provide the infrastructure for industrial and agricultural development, illustrate this ekistic policy model in the context of central command planning.

These new communities which provide their own comprehensive public amenities, equipment, collective administration, and community services (employment and production facilities, health care, educational services, housing, culture and recreation) are created in rural areas as developments organised around agricultural growth projects. New communities are also created in the periphery of urbanised areas as industrialisation projects designed to alleviate urban congestion and housing shortages without destroying existing urban facilities in order to create space for new structures. Within the context of a national central plan these new communities minimise objective differences in living conditions, services, and social development between urban and rural areas; meet the need for increased housing density in order to free land for agriculture; provide human settlement functions and housing for the labour force needed in regional development projects including economic production; simplify the delivery of basic social development services throughout the country; and overcome severe housing shortages with efficient, low-cost developments which provide upgraded community services.

The "Alamar" new community (15 miles east of Havana, built on 10 square kilometres) is designed for a population of 150,000 people; at present, however, there are approximately 50,000 inhabitants.

Alamar is divided into three basic districts surrounded by three wide, higher-speed streets. Each district is separated into micro-districts which provide residential and community service functions. There are five to six micro-districts within the larger district. Every micro-district has a population of 5,000 to 6,000 inhabitants. Thus, in ekistic terms, Alamar includes Communities Class V, IV and III.

Alamar's plan provides for one-direction growth (eastward) along the sea shore and 850 buildings (32,000 housing units). Approximately 700 buildings have already been completed (12,000 housing units). Each building is 4 and 5 stories high. Future plans call for buildings 12 to 18 stories high. Alamar also provides 13 day care centres, 8 semiboarding schools, 1 university, 5 supermarkets, 1 amphitheatre, 1 movie theatre, 1 bus station, 2 candy factories, 1 button factory, the largest laundry facility in Latin America, 1 furniture factory, 2 textile factories, 2 centralised kitchens (which prepares food for the schools, for day care centres, for the various work places and shops in the district), 2 polyclinics which serve approximately 32,000 outpatients each, and 3 swimming pools.

Alamar's housing is grouped according to occupation, not income. The members of a brigade which build a housing unit have the right to live in that unit; consequently, neighbourhoods are originally groupings of residents with similar occupations. In market countries housing selection and neighbourhood residential mixes follow income lines so that low-income families are usually located in one neighbourhood while higher income residents segregate in another location, a policy which results in economic and class segregation over space.

Rural-Urban Unbalanced Growth and Inequities

Another basic realisation that emerges from the comparative analysis of human settlements between the two groups is that nonmarket ekistic policy and practice serve the overall national objective of equitable territorial distribution of goods and services over space. Through the creation of new human settlements, employment and community facilities in rural areas, nonmarket countries, like China and Cuba, provide a more balanced rural-urban growth; have slowed down the inmigration trend of population movement from rural to urban areas which asphyxiates capital cities of Latin American, Asian and African countries. They have reduced considerably the inequities of goods and services between urban and rural areas. In the U.S.A. and other market countries, particularly those of the less-developed world, however, there is a stark contrast and inequity between urban and rural areas, between declining inner-city neighbourhoods (ghettos) and affluent suburbs, between large metropolitan cities which grow and those small cities which do not.

For example, Havana had 1.5 million population just before the Revolution of 1959, and was the only large human settlement in Cuba in the midst of small towns and villages - a situation similar to Santiago de Chile, Caracas, Bogota and Lima. By 1982 this urbanisation trend in Cuba changed dramatically as a result of an ekistic policy which emphasised the development and growth of small and middle size human settlements. The same policy applied to housing and community facilities including employment in the rural areas rather than in hydrocephalic Havana. Thus, Havana's rate of population growth and the associated urban sprawl was contained compared to pre-Revolutionary growth rates and to the growth rates of similar cities in Latin America. Havana grew to have a 1.9 million population (roughly 20-25 per cent increase) while Santiago de Chile, Caracas, Bogota and Lima have now close to 5 or 6 million population (roughly a 300-400 per cent increase rate). In these Latin American capitals, close to half the population live in squatter quarters while there are practically no squatters or slums of this type in Havana or in other areas of Cuba and China.

Cuba has reversed the urban-rural growth rate. Cuban cities in the provinces and rural areas grew at a faster rate than Havana, and most provincial capitals grew by more than 100 per cent. During the last ten years, moreover, Havana grew at an annual rate approximately .7 per cent, compared to a world urban population growth of roughly 4 per cent.

Mass Participation and Self-Government

A striking characteristic of nonmarket ekistic policy and practice is also the mass participation of people in the ekistic process. The "neighbourhood committees" in the People's Republic of China, for example, participate actively in the ekistic planning of their area and approve decisions regarding types and designs of dwellings, densities, community facilities, space organisation or dwelling distribution to residents. They also provide the administrative leadership and guidelines regarding maintenance, beautification and safety. Mass organisations like the Federation of Chinese Women, youth organisations and voluntary brigades of retired persons also participate in decision making processes regarding ekistic policy and practice at the neighbourhood level. Neighbourhood committees and mass organisations also provide input into their municipal and provincial councils regarding human settlement policy at the city, regional and national levels. In Cuba, like in China, active upward mass participation in ekistic planning is a fundamental feature of the central national development plan. In Cuba, for example, the Physical Planning Institute is part of the JUCEPLAN - the National Central Planning Board which guides the preparation and execution of one and five-year plans for the development of Cuba. The Physical Planning Institute, with its branches at provincial and municipal levels, collaborates with mass organisations at these levels in the formulation of recommendations to the JUCEPLAN.

This upward participation, as well as the horizontal mass participation in ekistic planning at every level, in collaboration with mass organisations, provide formidable multiple channels of self-governing and collaborative decision making. Housing policy in both countries, for example, is formulated and implemented not only by governments at various levels, but also by mass organisations, people's cooperatives, and by families through self-help projects. Construction brigades of volunteers, mass organisations, enterprises, and families, construct most of the housing units. The government provides the land, the plans, materials, tools and technical advise for families or cooperatives that decide to build their own housing, particularly in the rural areas. In small human settlements in Cuba like the Castillo de Jagua or Tablon, housing construction was undertaken by the residents on a voluntary basis. Master plans are prepared by the government and approved by the residents themselves.

Of all social services in nonmarket countries, housing is, relatively speaking, less spectacularly developed than health and education. Yet significant improvements have been made through decentralisation and mass participation of the population. As early as 1972, 75 per cent of all families in Cuba owned their own home, 10 per cent were still buying, 8 per cent paid rent, and 6 per cent were exempted of paying rent. Before the Cuban revolution approximately 60 per cent of housing was rental. Rent in both Cuba and China is now set at 7-10 per cent of family income, depending upon the participation of the family in the building process. Low-income families pay no rent. This contrasts sharply with ekistic practice in the U.S.A. where low income families pay 40-50 per cent of their income for housing (though 25 per cent of income is supposed to be the norm), and more than 75 per cent of Americans are priced out of the market for new, unattached dwellings (the cost being too high in relation to the average national family income).

In spite of the severe housing shortages that still exist in China and Cuba, and in spite of the moderate quality of housing compared to American housing, there are no homeless persons and one sees no slums.

CONCLUSIONS

The over-arching conclusion of this comparative analysis is that the human settlement crisis is a derived problem. Its cause is more intimately intertwined with ideological rather than technological of spatial constraints (Figure 1). Ekistic practice goes hand-in-hand with building participatory control of the socioeconomic forces which impact the structure and functions of human settlements.

Nonmarket territorial development is characterised by: (1) a firm pro-rural development and anti-urbanisation position in order to develop a more even, equitable distribution of population, resources and living standards; (2) a transformation of the ekistic planning process into a major vehicle of central national development; and (3) an infusion of mass participation and direct control by local population into ekistic policy and practice. As a result, ekistic planning in China and Cuba resulted in new style industrial-agricultural settlements like Taching in China and Alamar in Cuba. The development of the countryside has been directed towards optimal utilisation of all productive resources, including space. This pro-rural emphasis has favoured the diffusion of the industrialisation process and the development of small and medium sized towns and peripheral areas rather than a concentration in a few large cities.

To come to grips with human settlements policy and practice forces us to confront every ideological and social policy issue of modern society; e.g. the role of government power and income distribution and the morality of economic markets. This is precisely what links together individuals, institutions, governments and space. Certainly to master the body of human settlement knowledge is to experience in our own intellectual and professional development the most central themes of social, economic, political and spatial thinking crystallised in the last 500 years.

REFERENCES

Cox, K.R. (ed.) (1978) *Urbanisation and Conflict in Market Societies*, Chicago: Maaroufa Press.

Gurley, F. (1970) "Maoist economic development: the new man in the new China," *Review of Radical Political Economies*, 2: 26-38.

Iatridis, D. (1978) "Market forces and distribution policies," *Ekistics*, 45: 346.

Mesa-Lago, C. (1981) *The Economy of Socialist Cuba*, Santa Fe: University of New Mexico Press.

Mingione, E. (1981) *Social Conflict and the City*, New York: St. Martin's Press.

Tabb, W.K. and Sawers, L. (1984) *Marxism and the Metropolis*, New York: Oxford University Press.

CHAPTER 10

UTOPIAN LANDSCAPES

J.G. Nelson
Department of Geography
and
School of Urban and Regional Planning
Faculty of Environmental Studies
University of Waterloo
Waterloo, Ontario

SOME BACKGROUND THOUGHTS

Some initial guiding remarks about the words utopia and landscape will be helpful. The idea of utopia has a rich and long history in western thought alone. Treatises on the subject by authors such as Lewis Mumford, Ian Tod and Michael Wheeler show that the concept with its many meanings and implications has been with us for at least 4,000 years.

According to Mumford, a philosopher:

> The word utopia stands in common usage for the ultimate in human folly or human hope - vain dreams of perfection in a Never-Never land or rational efforts to remake man's environment and his institution and even his own erring nature, so as to enrich the possibilities of the common life. Sir Thomas More, early (16th century) coiner of this word, was aware of both implications. Lest anyone else should miss them, he elaborated his paradox in a quatrain which, unfortunately, has sometimes been omitted from English translations of his *Utopia*, the book that at last gave a name to a much earlier series of efforts to picture ideal environments. More was a punster, in an age when the keenest minds delighted to play tricks with the language, and when it was not always wise to speak too plainly. In his little

> verse he explained that Utopia might refer to either the Greek "eutopia" which means the good place, or to "utopia" which means no place. (Mumford, 1966, p. 1)

According to Tod and Wheeler, an architect and sociologist, respectively:

> Ever since Adam and Eve were expelled from the Garden of Eden, people have dreamt of Utopia. For most it has been a comforting faraway ideal in the distant past, the distant future, or a distant corner of the earth. But for a great many others, it has been an epic voyage of discovery into the possible future - an inspiration and a guide to the transformations of reality. People have attempted to realise it by experiment, example, reform, revolution, divine inspiration, or simply by withdrawing into the wilderness. Some have even hoped to construct it in steel and stone or gold and glass. Others have just laughed the whole idea of utopia to scorn. (Tod and Wheeler, 1978, p. 9)

Something of the guiding power and meaning of the word - the concept - to the creative individual and to society is reflected in the musings of the Greek practical dreamer, architect, and planner Constantinos Doxiadis on the occasion of his 1966 Lecture-in-Residence at Trinity College, Hartford, Connecticut, U.S.A. In contemplating the rapid growth and increasing complexity of the urban areas of the time, Doxiadis wondered how a rise in population from five million to eighteen million in Rio de Janeiro was to be accommodated by the year 2000. How was the growing industrial population of Ghana to be housed and how was the Georgetown Waterfront in Washington, D.C., to be saved from the interlocking highways "which spoil the Potomac landscape?"

> These are the cities, I said to myself, which we are supposed to ameliorate by adding new buildings and more modern highways and what is the result? We turn them into bad places - into dystopias. We are certainly not successful! What is wrong with us? Here is reality and here are our dreams - why don't they lead us anywhere? And then I came to the realisation that they are not properly connected because reality and dreams move on different planes and at different scales or speeds. What we need is a place where the dreams can meet with reality, the place which can satisfy

> the dreamer, be accepted by the scientist, and someday be built by the builder, the city which will be in place - the *entopia*.
>
> Sitting later alone, I came to the conclusion that this is the problem of humanity today; it builds cities which are bad, the *dystopias*, it dreams of cities for which there is no-place, the *utopias*; while it needs good cities for which there will be a place, the *entopias*. (Doxiadis, 1974, (3rd ed.), pp. XI-XII)

In his further musings on why things were less than ideal from the point of view of himself and many of his contemporaries, Doxiadis provides some reasons why utopias generally cannot be approximated, never mind achieved. The reasons are reflected in the voices that he hears from the past:

> "Why not build my Utopia?" I recognised Plato. Sir Thomas Moore stole my answer. "It is too small. You had better turn to my utopia." I then realised that my room was full of voices from the past and present representing dreams of the ideal city . . . philosophers, statesmen, architects, and many others - everyone looking at it in his own way. J.V. Andreae proposed Christianopolis, Etienne Corbet his Icaria, Edward Bellamy his America, Le Corbusier his Ville Radieuse, Frank Lloyd Wright his Broadacres City, and Aldous Huxley his Island. (Doxiadis, 1974, p. X)

Today many people are planning, are attempting to create Utopian processes if not Utopian features at scales beyond those of great cities - the Megalopoli - that preoccupied Doxiadis. And, as with earthly paradises, changes and plans in the past, the difficulties remain. Indeed they often arise and continue from the dreams and plans of the past - visions of utopias - incomplete and full of the unexpected, of surprises!

One of my main objectives is to encourage the bringing out of these uncertainties, difficulties, and challenges for careful thought and discussions in a manner that involves not only the professional and the so-called expert, but what some call the ordinary human being. The stress is on complexity and change. In this context it is important to remember that humans are capable of adjusting to increasingly impoverished environments and forgetting the quality left behind.

SOME DEFINITIONS

In making my points about perceptions, uncertainty, complexity, change, adaptation, and concern for the land and for people, I wish to discuss briefly what I see as four utopian landscapes: Banff National Park, Alberta-British Columbia; the Cypress Hills, Alberta-Saskatchewan-Montana; Paris, France; and Point Pelee, Rondeau, and Long Point Peninsulas on the north Lake Erie shore, Ontario. Each of these is utopian in its own right. But each respectively is also emblematic of a more general type of utopia: mountains and wilderness; the oasis in the drylands; the great city, the metropole, the centre of history, culture, and ideas; and the utopia for every man, the so-called vernacular or ordinary landscape.

In developing these utopian ideas and challenges the context is that of landscape, another word or concept that is ambiguous and yet full of opportunity. J.B. Jackson has laboured with the definitional problem as well as anyone; reaching back into Latin, Greek, and other languages for its roots and meaning (Jackson, 1984, pp. 5-8). One of his formulas has landscape as a composition of man-made spaces on the land. In this sense landscape is seen as a synthetic space, "a man-made system of spaces superimposed on the face of the land, functioning and evolving not according to natural laws but to serve a community . . . a landscape is thus a space deliberately created to speed up or slow down the process of nature . . . it represents man taking upon himself the role of time" (Jackson, 1984, p. 8).

Such a view of landscape seems accurate enough - to a point. But why is it necessary to limit meaning so much to the role of human beings? Are there not places at or near the surface of the earth where man has had relatively little influence and more is ascribable to so-called natural laws or processes? Such areas can still usefully be thought of as landscapes, with their landforms, vegetation, wildlife, and other characteristics. Thus, from a functional standpoint, the more man-made elements and activities often are closely linked to and depend upon natural or wilder elements nearby. This point will become clearer in the examples to be discussed later.

Jackson goes further toward the foregoing ideas in later defining landscape as, "a composition of man-made or man-modified spaces to serve as infrastructure or background for our collective existence" (Jackson, 1984, p. 18). In this broader sense, the word, landscape, is still short of a meaning that would relatively easily embrace the Arctic, the Kalahari, or other rather isolated parts of the world, where natural forces seemingly reign supreme, with relatively little obvious sign of man, although, as we will see later, different things may have been there before and been directly or indirectly affected by humans.

In this wide sense landscapes seem best thought of as the expression at or near the earth's surface of the collective, interacting network of natural, cultural, current and historic processes and features that are frequently referred to as the ecosystem. Or, to put it another way, landscape in an intersection with the ecosystem at some past, present, or future plane in time.

Another point that emerges from Jackson's writings is that the word landscape is not very precise as to scale or size. It can be used to refer to parts of villages, towns, or cities, or small rural areas. On the other hand, Jackson does use the word to refer to the agricultural settlements and other patterns that extend over thousands of square miles in the American West.

In thinking about the character, quality, or nature of landscapes - of how to read, perceive, and contemplate landscapes like those to be discussed shortly - some guiding concepts, ideas, or criteria of the geographer D.W. Meinig come to mind. In presenting these criteria or ways of thinking about landscapes, Meinig notes that the term and its implications deserve "the broad attention that only ordinary language allows" (Meinig, 1979, p. 34). In other words, thinking about landscape and its implications for nature and human beings is too important to be left to the professional.

In this context my mind goes back to a quote observed in a forestry museum in the Adirondacks, New York, in summer, 1986. It is attributed to a mid-nineteenth century woodsman and guide and reflects what must not uncommonly have been the thoughts of people who work closely with and rely upon nature.

To quote:

> It is an interesting study to look into things as they were so long aback and see what animals, birds, fishes and such things then existed; to know what of them have been pushed entirely out of the world, and what of them were left and to understand what changes white men and tame life all around them have worked on 'em. (Tucker, a guide to S.H. Hammond, Adirondacks, 1954)

SOME LANDSCAPE CRITERIA

Meinig presents the following criteria or organising ideas to make sense of what we perceive in landscape:

1. *Landscape as nature*: by this he refers basically to ways of perceiving the context in which humans live, the identification of plants, animals, soils, and the like, notions of nature as a stage, or as primary, fundamental, or enduring, or pristine, wild, primeval, or even bucolic or spoiled.
2. *Landscape as habitat*: by this he refers basically to the view that "every piece of the earth is the home of man." Man works at a relationship with nature; man domesticates and uses the earth.
3. *Landscape as artifact*: by this he refers to evidence of human effects upon nature, effects that some view as so extensive and intensive that the whole landscape has become an artifact of one sort or another.
4. *Landscape as system*: by this he refers basically to the workings and interactions of the hydrological, chemical, and other great cycles as well as more localised processes such as erosion and deposition which are often affected greatly by agriculture, manufacturing, or other cultural or human processes, in a word, the ecosystem. The short- and long-term workings of nature are important in this context.
5. *Landscape as problem*: by this he refers basically to the evocation of concern or alarm; the landscape is seen as a mirror of the ills of society and cries out for change; hence an interest in research, planning, design, and management.
6. *Landscape as resource or as wealth*: by this he basically refers to the economic view of landscape as a set of plants, floods, or other elements and processes that cost or benefit human beings.
7. *Landscape as ideology*: by this he refers basically to the perceptions, attitudes, and values that people associate with certain types of landscapes; factories and industrial areas are ugly from one perspective, or manifestations of economic or other power of a city, region, or nation from another perspective.
8. *Landscape as history*: by this he basically means that the landscape can be explained through the written or human record and through natural history, geology, biology, and other sciences. Every landscape is an accumulation. The past endures.

9. *Landscape as place*: by this he refers basically to landscape in the circumscribed sense, as a locality, "an individual piece in the infinitely varied mosaic of the earth."
10. *Landscape as aesthetic*: by this he basically means artistic qualities, recognising that the valuing of these varies among individuals and groups. Landscape is scenery, with beauty and other attributes that can be recognised. Meinig does not say so explicitly, but landscape also is more than visual; it is sounds and smells as well (Meinig, 1979, pp. 33-48).

Three other criteria or ways of thinking about landscape are as *symbol*, as *biography*, and as *vernacular*. By symbol is meant that certain landscapes or parts of landscapes bear unusually high value as indicative of some event or process at work, past, present, future. The Plains of Abraham in Quebec, the Hot Springs of Yellowstone, or Niagara Falls are examples.

The biography of landscape refers to the people who have created certain effects or patterns on the land, either as individuals or as groups. "Capability Brown" is an example, an architect who had fundamental effects on the English countryside through estate design in the eighteenth century.

J.B. Jackson is a leading if not the leading proponent of the importance of the vernacular or ordinary landscape. He stresses the understanding and learning about people, society, and their effects that can be derived by careful observation of everyday landscapes, the freeway landscapes of Los Angeles, for example, or the brownstones of Baltimore, functioning artifacts of the nineteenth century.

As a final point before turning to a discussion of some examples of utopian landscapes and their implications, mention should be made of the challenges that landscape can offer through learning, through the increasing capacity and discernment of the human beholder. This has nowhere been better expressed than by William Hoskins in his classic, *The Making of the English Landscape* (1955). In discussing the way a landscape changes through changes in the mind and so in sensitivity, Hoskins uses the analogy of the symphony. In so doing he also conveys a message about the pleasure and challenge that landscape, observation, reading, and thinking presents to all of us.

> One may liken the English landscape . . . to a symphony, which it is possible to enjoy as an architectural mass of sound, without being able to analyse it in detail or to see the logical development of its structure. The enjoyment may be real, but it is limited in scope and in the last resort vaguely

> diffused in emotion. But if instead of hearing merely a symphonic mass of sound, we are able to isolate the themes as they enter, to see how one by one they are intricately woven together and by what magic new harmonies are produced, perceive the manifold subtle variations on a single theme, however disguised it may be, then the total effect is immeasurably enhanced. So it is with landscapes of the historic depth and physical variety that England shows almost everywhere. Only when we know all the themes and harmonies can we begin to appreciate its full beauty and discover in it new subtleties every time we visit it . . . This book is . . . an attempt to study the development of the English landscape much as though it were a piece of music, or a series of compositions of varying magnitude, in order that we may understand the logic that lies behind the beautiful whole. (Hoskins, 1955, p. 20; quoted in full in Meinig, 1979, pp. 196-97)

FOUR UTOPIAN LANDSCAPES

In beginning my brief discussion of four utopian landscapes, please note that thinking is built around the basic ideas mentioned previously, that is, perceptions, uncertainties, complexity, change, adaptation, and concern for the land and for people. Thought is also organised around Meinig's guiding concepts or criteria for reading landscapes, that is landscape as nature, habitat, artifact, system - and so forth - but without necessarily referring to each criteria in evey case, although that would be possible, given sufficient time and space.

Banff National Park encompasses about 3,000 square miles of the Canadian Rocky Mountains. As a high, largely forested landscape, with many isolated peaks, valleys, and upland lakes, as well as patches of grassland and savannah, this mountainous terrain is perceived by many as a place to visit someday, a distant source of vicarious pleasure, a goal of ultimate fulfillment. It is a place to walk and climb in wild, unspoiled valleys where one can look over the vistas below; see the sun on nearby valley slopes; watch ice and snow glisten on the high far-away peaks; regard the distant flat mottled brown, green and yellow plains stretching far to the east.

Yet such mountains were not always seen by man as attractive, desirable environments in which to commune with nature, wilderness, the primeval, and God. There are also those who visit the mountains more regularly, know them more intimately, but often still hold them in awe. They hike, climb, ski, camp in, drive through, and otherwise use them, and often wish them to be planned and managed in such a way as to stress their utility to human beings.

Thus, the grizzly is seen on the one hand as a great creature of the wild mountains, a key element in this utopia, whose loss would seriously damage the image, do violence to nature, history, the system. On the other hand the big bear is seen as a brute, unpredictable, and dangerous, best confined to the nether regions or eliminated as threat to man and his enjoyment of the ultimate mountain playground.

Between these two extreme poles of thought, Banff National Park hangs - blowing in the wind: a landscape subject to varying views of the utopian ideal and so to the political process, the management plan, public participation, and the technician's hammer. It is, of course, a landscape that has changed much over the years in part for reasons independent of man, for example, climatic change and the retreat of glaciers, and in part as a result of the changing activities and ideas of man. A good example in this regard is the variation in vegetation arising from clearing, lumbering, mining, and other activities in the late nineteenth century pioneer days and from the ensuing regrowth of lodgepole pine forests because of the fire control and other protectionist policies introduced with the national park idea.

The Cypress Hills are not nearly so well known as Banff National Park but they represent another grand example of a type of utopian landscape - the oasis in the desert. The Cypress Hills are located in the southern Canadian plains, astride the Alberta-Saskatchewan border, with their slopes running into Montana and the United States. The Cypress Hills are the dark shadow observed by tourists to the south of Highway 1, the Trans Canada, near Medicine Hat, Alberta. Knowledge of these hills is as hazy in the minds of most of these observers as is their vague dark outline from the road.

Yet in places these Hills rise more than 400 metres above the surrounding drylands, clothed high up with forests of spruce and pine or lower down with parkland or savannah, mosaics of poplar, aspen, grasses, and the colourful flowers of spring. Water also is relatively plentiful in these uplands, caught in ponds, pools, natural and man-made lakes, and streams that carry life to the ranches, dryland farms, and irrigation systems around the Hills.

Like other drylands of the world those near the Cypress Hills have changing images. At times they are perceived to be much drier than at others. Rain and snow have varied through the decades and the centuries and will continue to do so for natural reasons. Perceptions of aridity have also changed among human beholders, in part because of real changes in aridity and in part because of varying visions of dryness provoked through previous experiences, the things the beholders were born and raised with, as well as their state of mind.

Thus, in July, 1859, the English explorer, John Palliser, who had been wandering through the Great Plains for many weeks, came to camp at the base of the Cypress Hills. Some of the feelings of Palliser and his colleagues are expressed in the journal entry for 28 July:

> The Cypress Hills formed indeed a great contrast to the level country through which we have been travelling; they are covered with timber, much of which is very valuable for building purposes, the soil is rich, and the supply of water abundant. These hills are a perfect oasis in the desert we have travelled . . . (Nelson, 1973, p. 104)

> On the twenty-ninth Palliser and his party entered: . . . "a magnificent valley" running through the "heart of the Cypress Mountains" and shedding "waters into the Missouri and into the Saskatchewan." Here they were "well supplied with wood, water and grass, a rare combination of happy circumstances" in their season's exploration. They also found the hills well stocked with wildlife. Grizzly, elk, and bison all seemed to be common. A "considerable number" of buffalo were killed on August 1 and the party spent that evening and the next day making pemmican. (Nelson, 1973, p. 104)

Palliser and his men were among the last to see the Cypress Hills ecosystem as it was prior to the arrival of the European with his technology, and his interest in trade and commerce. In the next twenty years the bison, beaver, and other elements of the system became resources to be exploited for sale in external markets and/or obstacles or hazards to ranching or other forms of settlement. Hence their severe reduction or elimination and replacement by cattle or other exotics from distant lands. The wolf, the grizzly bear, the bison, the human nomad, are no longer to be seen in the Hills where the landscape is dominated by irrigators, cattlemen, the recreationists who generally have settled into rough harmony with the deer, the fur bearers, the hawks, and other wildlife remaining from the truncated pre-European ecosystem. But these uses are in uneasy balance with more intensive agriculture and the greater development of cottaging, recreation, and tourism - bringing new changes. For example, Environment Canada Parks (formerly Parks Canada) is reconstructing Fort Walsh and other late nineteenth century trading posts and improving access for what is hoped to the thousands of tourists of summer. The imagery is neat and white, well painted, and devoid of the mud, manure, and smells characteristic of early days - a somewhat ahistorical reconstruction, but what most tourists seem to want, or are perceived to want anyway!

For most of us a major utopian image is that of the great city, ordered yet accessible, built yet somewhat natural, engineered yet artistic, historic yet modern, cultured yet democratic, comprehensible yet rather mysterious. Complexity and uncertainty mark such attributes, such characteristics. A variety of spectra intersect at shifting points in space, largely in accordance with history and changes in perception, taste, economy, and technique. Where then is the ideal environment, the utopia? Is it in the nature of the changing mosaic itself? Is there some essence - some part of the whole - that is basic to the utopian ideal?

Paris is for many people the ultimate city. For over 200 years it has been seen as a major if not the major centre of culture, learning, and discourse. It is said to have originated as an island settlement - a haven - some three hundred years after the birth of Christ and to have grown in accord with the styles and capabilities of the day. It has long been a centre of engineering and scientific experiment and development. Witness the great Exhibitions of the late nineteenth and early twentieth centuries as well as the Eiffel Tower and La Defense today. Art and Paris are synonomous, in painting, in sculpture, in gardens, in museums, in architecture. Paris is cultured yet democratic, as for example, in the cafes and in the incessant dialogue on life, events, and affairs, as well as on the politics of the quarters of the city, the country, and the world.

Paris is comprehensible yet mysterious, as befits a large metropolis, a symbol of a country, of a nation in the world. Its mystery lies in its size, its many districts, and backways, the varied building styles of a thousand years.

Much of its landscape biography is known in the works of kings such as Louis XIV or architects and engineers such as the nineteenth century builder, engineer, and architect Hausman. But much is unknown both in terms of creation, character, and origins. Knowledge of the geography and biography of Paris is the work of a lifetime, or more.

The essence of Paris as a city and of the great city as Utopia does seem to be the mosaic, largely comprehensible as to history, geography, architecture, style, dynamism, and overall character, but unknown and alluring, full of surprises, often surprises of recognition. For me the essence of Paris is global commerce and ways of life simmering in an architectural and social framework that is essentially French and European. Yet Paris is a dynamic meeting point of the world, centred in the old city, the river, and the islands. The architecture is certainly European, but reflects many influences, styles, and images from Islam and the East or other distant places. A long-standing question has been the appropriateness of new designs and proposals. Yet today the Left Bank, the nineteenth century apartments and boulevards, the Eiffel Tower and La Defence mix more or less compatibly together! They provide another essence of the utopian city: ordered diversity, difference, and stimulation.

The last example of a Utopian landscape which I wish to consider is what has been referred to earlier as Every-man's Utopia. In this case I am referring to those often little known, too familiar, and neglected landscapes found nearby. In the glow of information on other places, their qualities and their interest are often unappreciated or ignored. They are perceived as ordinary.

An example close to Waterloo, Ontario, is the North Lake Erie Shore and particularly the areas around the Three Peninsulas located along that coast: Point Pelee, Rondeau, and Long Point. From a natural, habitat, systems, and aesthetic standpoint, these peninsulas are unusually interesting and noteworthy. They consist of long strips of sand dunes and beaches, bordered with large expanses of marshes, wetlands, ponds, inlets, and the lake. They are also islands of forests, shrubs, and grasses amid surrounding polders, fields of corn and other crops, cottages, recreation, fishing, and industry. They are pressed on all sides and have been for more than a century.

These peninsular landscapes reflect much that is perceived as valuable from the past such as the attractive nineteenth century houses, the winding country roads, the historic fishing facilities, and the old boats and barns. These artifacts lead to knowledge of the rise and fall of various kinds of fishing, the United Empire Loyalists, the crash of the Great Lakes ecosystem, the creation of a rather harmonious agricultural landscape, bucolic, pastoral, and rich in the English notion of the countryside.

The Three Peninsulas also reflect the darker side of the Utopian image. They reveal the attempts to create economies, settlements, and ways of life to enrich and enhance human well being, attempts that were, however, built on incomplete understanding of the ecosystem. Whether these attempts are dreams and follies or unsuccessful attempts at seemingly rational development - limited albeit by unknown surprises - depends on your point of view.

A key question is how reasonable you think it is that certain experiences should be remembered by man and used to guide future action. An outstanding example in this respect is the attempt to build cottages and other structures on a permanent basis on the dunes, beaches, and coastal lands that make up the peninsulas and adjoining areas. But the lake goes up and down, some six feet or so over the decades, and has been known to do so for centuries. And the waves erode and transport the sediments and have been known to do so for decades. Yet attempts to build upon, stabilise, and conquer the shore have gone on steadily for decades as well, with much damage and many foregone dreams thereby.

Today the lake levels are very high and much destruction is taking place, in an atmosphere in which awareness of the essentially dynamic character of the peninsulas is growing. A different view of man in this ordinary utopia may yet develop in the not too distant future. This image may include large stretches of undeveloped beaches, boardwalks, and other devices to protect dunes, considerable open space along the shore, and less investment in breakwalls, revetments, groins, and other structures that break up the natural processes along the beach, work imperfectly, and detract from the utopian characteristics that many users see in the coast. Such images have been formulated at other times and places in Lake Erie and similar coastal environments but without changing the nature of the development processes that represent the dark side of the Utopian image.

SOME CONCLUSIONS

In concluding I wish to make a few simple basic points. Thinking about our surroundings and our past, present, and future effects upon them in utopian and landscape terms, although fraught with ambiguity, has some major advantages.

First, it makes us very aware of the fact that most new ideas are in fact old ones, thought of as ideal for human beings at some other time and place. The effects of these utopian ideas is therefore worth examining in terms of what they can tell us about current thoughts of the ideal future.

Second, by examining utopian landscapes, variations in what is thought to be good for humans and the land are revealed and much caution is acquired about proposals to "save the world" as these are made by particular professional or interest groups. The importance of dialogue, sharing, and attempts at consensus is underlined.

Third, and flowing from the foregoing, our general knowledge of utopias and utopian landscapes is largely western in origin; underlining the importance of being aware of and examining other people, cultures and perceptions and their views of the ideal, which can be so easily ignored, downplayed, or destroyed by others. Such efforts have been and are a marked facet of much of our thrust to development. Witness the native people of northern Canada today. Or think also about the Bushmen of the Kalahari, perceived as primitive in a largely worthless or waste land, meriting development by more civilised Caucasians since the sixteenth century. The Utopian quality of the Kalahari or Namibia drylands is, however, dimly discerned by reading the works of an informed observer such as Laurens Van der Post (1958), who reveals not only the remarkable accomplishments of the Bushmen as hunter, scholar of animals, plants, and the land, inventor, musician and dancer, but also some of the quality, mystery, grace, and beauty which Bushmen perceived in their homeland.

Fourth - and again following from the foregoing - thoughts of utopian landscapes tend to bring thinking about humans and nature more closely together. Such thinking makes us aware of the importance of protecting and husbanding not just the natural but also the human ecology of the places of the world. Both are intertwined, both are valuable, and in some utopian contexts, such as wilderness, threats to the human side may now merit the kind of attention devoted to nature in the past.

Finally, and here I am thinking very much of Doxiadis' ideas about the need for a place "where dreams can meet with reality" - this festival and this meeting is such a place, as is the University and the Faculty which hosts it. It is a place where attempts can be made to bring different dreams and realities together, to rub against one another, as we try to influence natural and cultural forces that ultimately are beyond design and complete anticipation. But such forces can, I think, be adapted to in such a way as to produce a nice balance between the two fundamental senses of utopia; first, the earthly paradise, with its soft, alluring, and somewhat hazy images of the good life; and second, the planned and ordered landscape with its practicality, tendency to uniformity and perhaps its lessened sense of challenge and joy.

Acknowledgements

This paper is based upon a talk given on the occasion of a fall 1986 Faculty of Environmental Studies academic festival on *The Unfinished Landscape*. I thank Sandy McLellan in particular for urging me to do this paper and playing a large part in arranging for the opportunity to present it. I also thank the Faculty and staff who helped organise the festival itself.

REFERENCES

Doxiadis, C.A. (1974) *Between Dystopia and Utopia*, (3rd. ed.), Athens: Publishing Centre.

Hoskins, W. (1955) *The Making of the English Landscape*, London: Hodder and Stoughton.

Jackson, J.B. (1984) *Discovering The Vernacular Landscape*, New Haven: Yale University Press.

Meinig, D.W. (ed.) (1979) *The Interpretation of Ordinary Landscapes: Geographical Essays*, Oxford and New York: Oxford University Press.

Mumford, L. (1966) *The Story of Utopias*, New York: Compass Books Edition, The Viking Press.

Nelson, J.G. (1973) *The Last Refuge*, Montreal: Harvest House.

Tod, I. and Wheeler, M. (1978) *Utopia*, New York: Crown Publishers, Inc.

Van der Post, L. (1958) *The Lost World of the Kalahari*, Harmondsworth, Middlesex, England: Penguin Books.

CHAPTER 11

FRONTIER SETTLEMENT AND HUMAN VALUES: A COMPARATIVE LOOK AT NORTH AMERICA AND SOUTH AFRICA

Leonard T. Guelke
Department of Geography
University of Waterloo
Waterloo, Ontario

The interests of a historical geographer would appear to be somewhat removed from a theme which includes abstract thoughts and concrete solutions. Yet notwithstanding its academic character historical geography is not without relevance for those seeking solutions to concrete problems. When geographers and planners attempt to solve problems they deal with people and their values. The basic values of a people are in turn historical creations. The experience of a people shape their values and set the limits on the kinds of solutions that can be implemented successfully. An understanding of the values of a people in a historical context is needed if envisaged solutions are to be firmly grounded on enduring human values of those who must make them work.

In this essay I am concerned to elucidate the relationship between historical experience of immigrant settlers and human values on the frontier. If frontiers were important in changing ways of life and creating new values, the length of time settlers were exposed to frontier conditions would presumably be an important element in determining the extent of that transformation. In particular, if frontiers provided the impetus for settlers to be more self-reliant and individualistic their disappearance could presumably weaken these frontier values. Thus an inquiry into the origins of values and conditions that sustained them is also an inquiry into the possible modern relevance of such values.

The meaning of frontier settlement in the context of European expansion overseas has been widely misunderstood by historians and historical geographers, because both the defenders and critics of the celebrated frontier hypothesis of Frederick Jackson Turner have lacked a clear definition of the word "frontier." A definition, by itself, is not necessarily crucial, but in the case of frontier studies the lack of an adequate definition has led to two essentially different phenomena being confused with each other. One of these phenomena has to do with the meaning of the frontier proper conceived of as a zone of interaction between European intruders and the autochthonous peoples (Lamar and Thompson, 1981, pp. 3-13). The other with the broader issue of the settlement of new lands. In this paper I confine my attention to the meaning of the frontier proper for European settlers in North America and South Africa. In limiting my study to the European side of the frontier, I do not wish to imply that it is the only side of the frontier worth considering. A complete frontier study would be one which examined both European and autochthonous peoples, and indeed historians with appropriate backgrounds are now, and have for some time been, rewriting frontier history from the points of view of native peoples (Axtell, 1978, pp. 110-44).

In the context of European expansion a frontier was created or opened with the first permanent European settlements on native land. The setting up of trading posts by Europeans in new lands or hunting in native territory are not in themselves considered frontier phenomena. These activities only become frontier ones when they are associated with advancing European settlement and the dispossession of native peoples of their lands. The frontier phase of settlement ended, or closed, with the elimination of native people as active contenders for the land and resources of the areas they once controlled. This development typically occurred when the resources of an area were no longer capable of supporting both the original inhabitants and the increasing number of white settlers who sought to make a living from the land. The closing of a frontier was not generally associated with large numbers, because frontier settlers used resources on an extensive scale and needed large areas of land to make a living for themselves. The closing of the frontier often marked the beginning of commercial development, because the increase in population made the more intensive and specialised use of resources both possible and necessary.

In both South Africa and North America European expansion created frontier areas or zones where settlers of predominately European origin vied with autochthonous peoples for land and resources. In this unequal contest, well-armed European colonists rapidly expanded their settlements at the expense of the native peoples. In North America European settlers moved westwards across the continent from settlements established along the Atlantic Seaboard. In South Africa an initial tiny settlement made at Table Bay in 1652 was expanded to the east and north and within two centuries much of Southern Africa south of the Limpopo was controlled by European settlers. In North America the scale of the movement was vast and frontier conditions were rapidly replaced by stable, new societies. In South Africa, where the number of European settlers was small, many people were exposed to frontier conditions for protracted periods. The South Africa case is of great potential value for testing the validity of some of the generalisations about the meaning of the frontier that have been put forward by historians of North America, because it permits one to examine the nature and effects of the frontier in a situation much less complex than that of North America.

There are many questions that one might ask about the meaning of the frontier in the context of European expansion. In this paper my attention will be confined to three distinct, but interrelated questions, which seem to lend themselves to analysis from a historical geographical point of view. First, what did the frontier offer the people who settled it? This question has to do with the motives of frontier settlers and particularly whether these motives were of an economic, social or political character. Second, what did frontier conditions mean for the people who comprised settler societies? Third, what were the ramifications of the frontier experience on the societies which developed in its wake? I will now consider each of these questions in turn.

FRONTIER EXPANSION

The rapidity with which much of North America and South Africa was settled by European colonists suggests that the frontier was perceived as an area of opportunity, in spite of its isolation and dangers. Moreover, the attractiveness of the frontier seems to have been unrelated to its immediate or even long-term commercial potential. In North America many settlers moved into areas such as the Piedmont, the Appalachian Mountains and Kentucky before transportation facilities were developed and lived in general isolation of the commercial economy for decades. In South Africa the trekboers put hundreds of miles of open veld between themselves and their only market at Cape Town. This rapid movement of settlers into the interior was not anticipated either by the British in North America nor the Dutch in South Africa, and its meaning remains a vital question for modern scholars of European expansion and frontier settlement.

Although the basic ingredients of frontier settlement were people and land, an explanation of frontier expansion based on population pressure and the availability of "free" land is inadequate. The fact to be explained is not the phenomenon of a growing population, but rather why that population, or a portion of it, moved to the frontier rather than stay put, or perhaps, seek a livelihood from the sea. A satisfactory explanation must be able to show that the frontier offered certain people, in specific economic and social circumstances, positive advantages over the other alternatives that might have presented themselves. The availability of "free" land by itself, is not necessarily correctly considered a positive advantage. In the first place, frontier land was not free; it had to be taken from native peoples. In the second place, free land is worthless unless it is capable of being effectively used. The land of Alaska (even if free) is not an answer to unemployment, because the unemployed of New York City would generally not be in a position to make a living from it. The real question of frontier settlement to be answered, therefore, concerns the kind of living it offered European settlers and the advantages that that living had over the available, or, more precisely, the perceived alternatives.

In both North America and South Africa frontier settlement was characterised by a large degree of subsistence economic activity. Indeed, had it not been possible for settlers to support themselves in large measure from such activities it is unlikely frontier expansion could have occurred as rapidly as it did. Although practically all frontier scholars acknowledge the importance of subsistence activity, particularly in the early years of settlement, the interesting question is whether this activity was an early phase in the commercial development of an area or a phenomenon in its own right. In South Africa and certain isolated areas of North America, where commercial economies developed slowly, it is possible to examine the question in a setting free of the complications that are associated with rapid commercial developments that occurred in much of North America. If it could be shown that, even in remote and isolated areas, commercial considerations were paramount any need of a separate explanation of frontier settlement would have been eliminated.

One school of thought has maintained that, even in areas largely isolated from outside markets, commercial considerations were never far from the settlers' minds (Mitchell, 1977). In support of this position it has been pointed out, notably by Neumark (1957) in the case of South Africa, that however remote the frontier, settlers always maintained some commercial ties with the outside world. The crucial issue here is not the existence of commercial ties themselves, but rather whether such ties were profitable from a commercial point of view. If the commercial ties were not profitable, they can scarcely be said to have provided a commercial motive for settlement. The evidence is overwhelming that in isolated areas of North America and practically all of South Africa the trade maintained by settlers of the frontier with the outside world was, from a commercial point of view, unprofitable. Gray (1932, 1, p. 451) has made the following accurate observation with respect to the Southern United States:

> What if the production and hauling to market of hogsheads of tobacco did cost twice the labour incurred by commercial planters, provided the self-sufficing farmer could obtain a gun, ammunition, kettles, and medicines, which he could not produce at all or only by excessive labour and trouble? He was not inclined to balance the money return from the product against the money cost of producing it, but rather the labour of producing and marketing the money crop against the utility of the commodities he was thereby enabled to buy.

John Barrow, an English observer of late eighteenth century South Africa made a similar point:

> The distance (separating the frontier from Cape Town) is a serious inconvenience to the farmer . . . if he can contrive to get together a wagon load or two of butter or soap, to carry with him to Cape Town once a year, or once in two years, in exchange for clothing, brandy, coffee, a little tea and sugar and a few other luxuries, which his own district has not yet produced, he is perfectly satisfied. The consideration of profit is out of the question. A man who goes to Cape Town with a single wagon from the Sneuwberg must consume, at least, sixty days out and home. (Barrow, 1801, 11, p. 331)

Additional evidence against the commercial interpretation of frontier settlement in isolated areas of poor communications is provided by an examination of the rates of settlement and comparing them with the market for frontier produce. Frontier settlement proceeded regardless of the fluctuations in the market for the produce of the frontier. In South Africa, for example, new settlement continued at rates that showed no correlation with the fluctuations in the prices that meat was fetching in the market at Cape Town (Guelke, 1976, p. 40). In the early 1740s the market was particularly depressed, yet the settlement of new frontier areas showed no evidence of decline. Neither did difficulties of transportation and poor markets deter settlers of remote areas in North America. However, in the North American context the commercial settlement of many inland areas was rendered possible almost from the beginning of the frontier by the existence of navigable rivers. There is no problem of explanation here; the problem is posed by the subsistence settlement of areas which lacked prospects of commercial development - not for a few years but for decades and even scores of years.

The nature of frontier expansion becomes more rather than less puzzling if commercialism is rejected as the major factor behind much frontier settlement. Why were European colonists eager to take up and settle land without immediate commercial value, often in defiance of government decrees and without government protection? In terms of conventional economic theory a subsistence economy represents a step backwards from a commercial one, because the subsistence producer sacrifices the advantages that arise from specialisation based upon a division of labour. On the acceptance of this theory the frontier would not have been a desirable place to settle, and would have attracted only those who were in desperate economic straits. However, the rapidity with which the frontier was expanded in both North America and South Africa makes any interpretation based upon the notion of reluctant settlers untenable. This economic argument seems to lead to a rejection of economic factors as an explanation of frontier expansion. There is, apparently, no alternative but to accept the importance of social and political factors, and to interpret the frontier as a place for freedom seekers and individualists who sought to escape the burden of heavy taxation and the oppression of arbitrary government. The opportunity the frontier offered people in terms of measure of independence was likely a contributing factor behind frontier settlement, but it seems unlikely large numbers of people would have risked danger and isolation to assert their individuality *against* economic forces.

There is an alternative economic explanation of frontier expansion, which avoids the impasse reached in the previous paragraph. This explanation is based upon the rejection of validity of conventional economic theory in the special conditions that applied to frontier settlements. In these conditions, which involved a small population amidst abundant resources, it was possible for many settlers to support themselves with less effort than they were able to do as inhabitants of commercial regions which enjoyed the benefits of the division of labour. Boserup (1965, pp. 43-55) has shown, in her important work, *The Conditions of Agricultural Growth*, that in rural economies the amount of effort needed to support a family increases dramatically as the amount of land available diminishes. The end result of an increase in the population of a fixed land area is not necessarily a lower standard of living for all, because as the population increases economies based upon a division of labour can be realised. However, even in situations in which such economies are realised, a good living is not to be had without everyone working a lot harder than had been necessary in the days of abundant resources.

In support of her thesis Boserup cites a number of studies of African and Asian subsistence agriculture which are of great interest. In a study of the Bemba in Zambia, Richards (1939, pp. 393-95) estimated that, on average, a family was able to produce all the food they required with about three to five hours work per day. This work included clearing the land of trees and bushes, preparing it for cultivation and sowing and harvesting the crops. When their fields were exhausted the Bemba moved to a new area, and, indeed, the success of their agriculture was dependent on a low man/land ratio, which allowed old cultivated areas a chance to recover their fertility. Similar conditions to those of the Bemba have been described for others, notably de Schlippe's work on the Zande (de Schlippe, 1956, p. 168). The ease with which Zande subsistence cultivators were able to make a living enabled them to devote much of their free time to beer drinking and hunting. When Europeans observed the relaxed lifestyles of such people they were apt to dub them "lazy" and colonial administrators had their time cut out thinking of ways to make them "useful" workers.

There are important parallels between the African subsistence cultivators described above and many of the frontier settlers of North America and South Africa. Although these settlers were apt to operate as individuals on their own lands (that is, these settlements lack the communal aspect of African subsistence economies) they enjoyed the basic economic advantages of other subsistence cultivators and more. The frontiersmen of North America and South Africa were equipped with rifles, which made it easy for them to exploit the abundant game resources of frontier areas and lessened the dependence on cultivated crops. The raising of such crops, however, was not a time-consuming or difficult task. Fields were seldom cleared of the stumps and the crops were sown extensively in "fields" which were abandoned as they lost their fertility. The time consuming methods of the farmer intent on increasing yields, such as manuring and weeding, were not necessary. A frontier living was to be had without effort by settlers lacking capital, because frontier life was premised on the direct exploitation of an area's natural resources. A contemporary observer in the late eighteenth century described the prospects of Kentucky as follows: "A new country called Kaintuckey . . . is recon'd the finest country in the world affording almost all the necessities of life *spontaneously* (my italics)" (quoted in Wertenbaker, 1963, p. 148). This type of frontier would have been most attractive to labours and poorer settlers, who lacked the capital to maintain or create an independent livelihood in the settled regions. In 1717 Governor Spotswood of Virginia (quoted in Turner, 1961, p. 51) wrote:

> The inhabitants of our frontiers are composed generally of such as have been transported hither as servants, and, being out of their time, settle themselves where land is to be taken up and that will produce the necessarys of life with little labour.

The frontier, viewed from a resource perspective, would have had a positive attraction for many, especially poorer, inhabitants of the European settled areas of North America and South Africa. In South Africa it is unlikely that the frontier would have expanded at all had it not been possible for settlers to make a good living for themselves in the frontier regions, because commercial development often lagged many decades and even scores of years behind the first settlement of a region by European colonists (see exchange between Norton and Guelke, 1977, pp. 463-67). This was not the case in North America, where frontier conditions rapidly gave way to commercial developments. Yet even in North America many areas owed their first settlement to settlers who were more interested in resources than land. These people often bore the brunt of Indian hostility, and were apt to move to new areas with the advance of close settlement. The importance of this first noncommercial wave of settlement, which I have identified as being closely associated with the frontier, has possibly been underestimated in North America historiography, because of a lack of evidence. The people who followed them appropriated the term pioneer and left copious records of their achievements in the form of pioneer reminiscences. Such people would have not had much interest in praising those who might have preceded them. Yet the evidence that a subsistence phase of frontier settlement was typical of North America is to be found in many areas, particularly those areas, such as the Piedmont and Appalachians, which were slow to develop as commercial economies. In such areas the frontier is directly comparable to the frontier in South Africa. In many areas of North America, however, the frontier was a complex phenomenon and subsistence farmers by choice are found interspersed with reluctant subsistence farmers who represent the first wave of commercial development.

FRONTIER ECONOMY AND SOCIETY

In the preceding section I argued that the frontier offered European settlers a subsistence livelihood based upon abundant resources. These resources, however, remained abundant only so long as there were few settlers and native inhabitants. An increase in the settler population made the extensive system of resource exploitation less and less productive. The rise of the settler population made it necessary for frontier inhabitants to adapt to new conditions or to move on to new frontiers where they could continue their old ways of life. Only in areas which, for one reason or another, were unattractive to commercial exploitation, did frontier ways of life outlive the frontier. A fundamental fact of frontier existence, which made possible both the settler subsistence economy and the co-existence, which made possible both the peoples in one region was a generally low man/land ratio. Frontier populations were typically less than one or two persons per square mile. These low population densities, in turn, had important implications for the nature of the frontier economy and society (Table 1).

The frontier lacked hamlets and villages. In South Africa the extensive nature of settlement precluded the emergence of any service centres in the interior throughout the eighteenth century. There were no inns, no stores, no schools or medical services. The population was so sparse that it was simply not able to support such services. The situation was much the same in America, although the frontier phase was typically of a much shorter duration. In the absence of urban services the frontier inhabitants were thrown on their own resources. Travelling preachers and teachers catered to some of their needs, but in other areas frontier people developed handicrafts and folk remedies. Although individuals were unable to master a variety of tasks, they became expert in the skills needed to survive in the wilderness. If life lacked refinement it was compensated for by the ease with which a living, albeit a rough one, could be made.

Table 1

Some Interrelated Factors of Frontier Settlement

extensive agriculture	=	low population densities
low population densities	=	poor communications and services
poor communication and services	=	inadequate protection and control by governments; economic self reliance
economic self reliance	=	limited social interaction, isolation
isolation	=	independent and individualistic attitudes

The isolation of the frontier was exacerbated by a general lack of good communications. It was essentially uneconomic for the frontier population to build roads or bridges. In South Africa many decades elapsed before rivers were bridged or before wagon tracks gave way to roads. In such conditions, which are also to be found in many areas of North America, it is not surprising that governments had problems controlling frontier settlers, and that frontier settlers had little use for governments. As long as frontier conditions existed governments were unable to offer frontier inhabitants either services or even protection against attacks by native peoples.

The lack of services, paradoxically, made frontier inhabitants if anything less rather than more hard working. The general isolation worked against those who attempted improvements and favoured those who sought a living directly from the available resources. The marginal gains to be had from hard toil, of a commercial kind, and the ease of making a livelihood gave frontier people considerable amounts of leisure time. In South Africa travellers were apt to describe the trekboers as lazy. The Swedish visitor Sparrman described life of a recently settled frontier district.

> All the colonists who follow the grazing business, and particularly those at *Agter Bruntjes-hoogte*, lead an easy and pleasant life. One of these boors usually gets to his plough eight or ten of his fat, or rather pampered oxen; and it is hardly to be conceived, with what little trouble he gets into order a field of moderate size . . . So that, always sure of a rich harvest from a soil not yet worn out, which is ever responsive to the culture bestowed upon it, he may be almost said merely to amuse himself with the cultivation of it for the bread he wants for himself and his family; while many other husbandmen must sweat and toil themselves almost to death, both for what they use themselves, and for that which is consumed by others . . . (Sparrman, 1977, II, pp. 130-31)

Conditions in newly settled regions of America were much the same. The following description pertains to the back country of North Carolina:

> Nature holds out to them everything that can contribute to conveniency, or tempt to luxury, yet the inhabitants resist both, and if they can raise as much corn and pork, as to subsist them in the most slovenly manner, they ask no more; and as a very small portion of their time serves for that purpose, the rest is spent in sauntering thro' the woods with a gun or sitting under a rustick shade, drinking New England rum made into grog, the most shocking liquor you can imagine. (Lefler and Powell, 1973, p. 179)

The ease with which many were able to make a good living on the frontier has not been adequately appreciated. In his book on the American frontier, Rohrbough (1978) emphasises the difficulties of frontier living and the backbreaking toil it demanded. These descriptions are probably accurate if applied to the post-frontier phase of commercial settlement, which demanded investments of time and capital in the development of individual farms and a regional infrastructure. The frontier proper, however, did not demand backbreaking toil, and for those who were not concerned with the wider society it probably provided a rather satisfying and easy-going life. I suspect many frontier visitors would likely have been blinded to the real nature of frontier society, because it failed to fit their preconceived ideas about rural society, in much the same way that early European travellers in Africa misinterpreted African subsistence economies.

The celebrated Turnerian characteristic of frontier individualism was another important implication of frontier settlement. Apart from the needs of defence and travel, each frontier family and their dependents were able to produce almost all their requirements on their own. There were few economic incentives for people to cooperate with each other - for example, in the provision of roads or fences - and a fundamental economic reason for established frontier inhabitants to frown on new settlement, which threatened their way of life. In these circumstances individualism had a fertile ground in which to flourish, and so it did in both South Africa and North America. In South Africa individualism was pushed to extremes and it was with difficulty that any kind of cooperation on an extended basis was achieved (Lichtenstein, 1928, 1, p. 116). The Great Trek, for example, was punctuated by disagreements and bickering among leaders, who had difficulty in placing their personal interests behind the common good. In North America "frontier" individualism long survived in the Appalachian mountains, which were not strongly affected by post-frontier commercial settlement. In areas affected by commercial developments individualism was replaced by the tolerant and cooperative attitudes that are essential in a complex commercial society. The individualism that is so often thought of as a general characteristic of early America is largely a myth.

A rise in population was both a cause and an effect of the spread of commercialism. The early commercial economies which emerged to replace subsistence agriculture and hunting were solidly based on acquisitive values and the division of labour. Communications were improved. Towns grew up rapidly to supply the farmer with specialised services. Newspapers were founded and schools and even universities sprang up to serve a rapidly growing rural and small town population. The extension of settlement within a framework of commercialism preserved the essential values of hard work, tolerance and cooperation which are the hallmarks of the early American commercial economy. Individualism is here a luxury few can afford and democracy a good solution to choosing leaders in the absence of established or traditional rulers.

Although the commercial settlement of a new region, already secure from attack and with access to markets was a quite different phenomenon from early frontier settlement, it has frequently been confused with it. I would defend the interpretation of frontier put forward here on the ground that frontier settlement is associated with a distinctive social and economic order, which distinguishes it from later commercial developments. Indeed, where settlers were able to take steamers and trains to surveyed land in the Mid-West, and to be provided within a few years with all amenities of civilisation, one can safely deny that this was frontier settlement in any reasonable sense of this word. If such worthy "pioneers" later chose to relive these important episodes in their lives in grand memoirs, the historian of these times has special need to separate reality from legend.

RAMIFICATIONS OF THE FRONTIER

It is unlikely that Turner's frontier thesis would have become as important as it did had it not linked frontier settlement with the evolution of American democracy. The crucial questions to be considered are the ramifications of the frontier on North American and South African societies. In such an examination it is important to keep separate two questions, which I believe Turner confused. We must differentiate the effects of the frontier from the economic opportunities created by the land itself. For example, in Europe the Black Death created some of the conditions sometimes ascribed to the frontier. It would obviously be stretching the term frontier beyond any reasonable meaning to label the Black Death a frontier phenomenon. I will keep closely to the definition of frontier outlined above. This emphasis is not considered arbitrary; I would probably agree that the land, as opposed to the frontier, was important in shaping the nature of American society. Yet if land (in conjunction with other factors) was the key factor it seems right that it be recognised as such rather than being subsumed under the word "frontier" whose connotations are clearly more limited.

In my introduction, I suggested, the frontier closed with the elimination of autochthonous peoples as contenders for the resources of an area. This phase typically occurred when it was no longer possible for all the frontier inhabitants to make a living from extensive cultivation, open range stock raising and hunting. The increasing settlement of an area by white settlers led to a shrinkage of open range, a decline in soil fertility and the disappearance of game. Once the native people had been eliminated the frontier inhabitants were faced with having to change their methods of farming or move to new areas of "unspoiled" wilderness. The ease with which frontier peoples were able to adjust to the post-frontier world was probably related to the length of time they had experienced frontier conditions.

The frontier population was always a small fraction of the total population of North America. Although the early frontier people performed as it were a service to those who followed them by eliminating the Indians as serious contenders for the land, they had little impact in shaping the values for those who settled after them. Indeed many independent frontiersmen, accustomed to living off the land and fighting Indians, had a painful adjustment and ended their lives as poverty stricken, marginal members of the main society. The plight of the Wetzel brothers of West Virginia is typical. These four brothers having spent a lifetime on the frontier petitioned the state for a pension claiming that they deserved it because they had killed thirty-eight Indians between them and that they were (partly as a result of their wounds) no longer able to work as farm labourers, which was the only occupation they knew (Rice, 1970, pp. 168-69).

In North America the frontier closed at an amazingly rapid rate. The existence of many navigable rivers allowed for the commercial penetration of the eastern parts of the continent. The prospects of commercial development were never far away and some regions escaped entirely the kind of frontier phase of settlement I described above. The advent of the railroad which linked new areas with older ones effectively ended frontier settlement. In a situation allowing for effective government control and easy access to markets a frontier could scarcely be said to exist. The end of frontier America generally occurred well before the Civil War, although frontier conditions survived until much later in isolated mountainous and Western areas.

The impact of the frontier was greatest in those areas which experienced slow economic growth such as Quebec (New France) and Appalachia. In these areas, where the population was not for one reason or another augmented from the outside, frontier ways of life were able to survive long after the frontier had disappeared. In these areas population pressure eventually made change imperative, but change came slowly and frontier values - such as noncommercial attitudes and easy-going work habits - survived long after they had outlived their usefulness. The people of Appalachia, and Quebec, for example, after decades if not centuries of isolation faced serious problems reintegrating themselves into the mainstream of North American life.

The experience of the white settlers of South Africa is typical, if anything, of the isolated areas of North America (Guelke, 1985, pp. 434-48). The early commercial development of South Africa was hampered by a lack of good communications, and the problem was not ameliorated at all by the presence of navigable rivers. Many generations of settlers were exposed to actual or frontier-type conditions. When increasing commercial opportunities began to develop in the early nineteenth century English-speaking immigrants were in a stronger position to take advantage of them. Not until the discovery of diamonds and gold was the isolation of interior South

Africa decisively terminated, but many boers were able to postpone coming to terms with the modern commercial world until the twentieth century.

The long frontier experience of the Afrikans-speaking, white population of South Africa has had important ramifications for the evolution of that country. In the process of commercialisation many poor quasi-subsistence farmers and their children were displaced from the land (Giliomee, 1981, pp. 76-119). Lacking modern skills these people were at a grave disadvantage in the job market of Southern Africa. The employers in the growing industries favoured black or other (Indian, Chinese) unskilled labour, because it was cheaper than white. In this situation the poor whites, the casualties of European frontier expansion, translated their political power into economic power. A series of apartheid-like measures were enacted by various white governments to give these people sheltered employment or provided opportunities for them to learn the skills they need in the modern economy. It would probably not be too much of an exaggeration to attribute the rigid apartheid policy of modern South Africa to the long frontier experience of its Afrikans-speaking white population.

To conclude, the frontier, was of crucial importance for South Africa and of marginal importance in America. We can see what America might have been had the frontier been of great importance by looking at the isolated societies of the Appalachians. The values of these mountain people would have been those of the majority rather than those of a quaint minority. Nevertheless, if the frontier is dismissed as a major factor in North American development, to what do we ascribe the emergence of such crucial characteristics of the society as democracy? This question cannot be answered in a sentence, but the answer it seems must be sought in the distribution of power. I have argued that the frontier people in North America were never a large enough group to exert a formative influence on that society in contrast to the situation which developed in South Africa. Whatever else might be said it seems to me a truism that societies are shaped by the powerful (as individuals or groups) and not by the weak. In America simply by virtue of their numbers alone the frontier people never became a powerful force in the larger society.

SUMMARY AND CONCLUSION

The lack of a clear definition of the word "frontier" has resulted in much avoidable confusion in the historiography of the frontier. At issue here is not a minor matter of precise definition, but rather a basic question concerning the failure of many frontier scholars to differentiate two quite distinct and empirically identifiable phenomena. These scholars, including Turner himself, failed to appreciate the crucial differences between the initial, largely subsistence phase of frontier settlement, associated with the wresting of land from native peoples, and the subsequent consolidation of settlement and the rise of commercial agriculture. Turner certainly identified the subsistence phase of frontier settlement (in citing a long passage from Peck's, *New Guide to the West*, Boston, 1837), but treated it as a part of a general frontier process, and failed to understand its real economic and social meaning (Turner, 1961, pp. 49-51).

In both South Africa and North America frontier expansion was promoted by settlers who sought a subsistence livelihood. In North America, however, the situation on the frontier is often confused by the rapidity with which commercial agriculture developed. Where railroads or canals made commercial agriculture possible from the beginnings of settlement one cannot really talk of a frontier phase of settlement at all. The frontier is essentially a phenomenon of pre-industrial technology. The advent of effective communications destroys subsistence agriculture and makes possible effective government control.

In South Africa the lack of commercial development associated with a rising white population permitted frontier and frontier-like conditions to endure for decades, if not centuries. In this respect the South African case is of great importance, because here it is possible to study the effects of the frontier in a situation uncomplicated by commercial developments, which so rapidly destroyed the frontier in most areas of North America.

The frontier is here conceived as a zone of interaction and conflict between European settlers and native peoples. This frontier was characterised by low population densities and a lack of communications with the outside world. For the settlers of the frontier the resources of the land had a greater value than the land itself (Turner, 1961, p. 49). The general isolation of frontier people from each other and the outside world militated against an exchange economy. The frontier people were thrown on their own resources in making a living and had generally to protect themselves from hostile native peoples. Under such circumstances it is surprising settlement occurred at all. That it did was largely the result of an economic paradox, which made subsistence living desirable for a small population in an "unspoiled" land. This "paradox," rather than a love of freedom, encouraged settlers, especially those lacking capital, to move to the frontier in spite of dangers and isolation, ahead and independently of those who were interested in its commercial development.

The lack of communications and an absence of small towns was characteristic of the economy and society of frontier areas. Each individual family or group was largely independent of its neighbours and performed practically all economic tasks on its own. The isolation of frontier families encourages self-reliance and individualism, because the common good was the good of the individual frontiersmen and his dependents. The sparse population made the provision of common services uneconomic and individuals were discouraged from developing their lands because of a lack of markets. In this environment democracy was largely irrelevant. Each individual could afford to be a law unto himself, except in matters pertaining to defence and, to a lesser extent, travel. The attitude to governments was, not surprisingly, generally negative. The authorities were unable to provide frontier people with adequate protection or services at a reasonable cost, because the people of the frontier were so dispersed.

When frontier conditions were replaced by commercial developments the frontier people had either to adapt or move if they were to avoid becoming destitute. In both North America and South Africa many whites appear to have been perpetually on the move, because they were not equipped for nonfrontier ways of life. Others attempted to adapt to new conditions. If they were successful they became indistinguishable from their commercially-oriented neighbours. If they were not successful they became fringe members of the main society - poverty stricken and largely forgotten.

The frontier was of minor importance in shaping North American society and civilisation. In the first place, because so few people were involved and, in the second place, because the rapid growth and spread of commercialism swept the frontier away before its effects could be widely felt. Only here and there in isolated areas lacking commercial prospects were frontier ways of life able to endure for long. In Appalachia, and to some extent in Quebec, frontier-type conditions survived for long enough to fashion distinctive, but atypical, American societies. In South Africa frontier conditions were experienced for decades by a substantial number of white settlers and their nonwhite servants. This frontier experience shaped eighteenth and nineteenth century South Africa and is of central importance for anyone who would understand the modern history of that country.

REFERENCES

Axtell, J. (1978) "The ethnohistory of early America: a review essay," *The William and Mary Quarterly*, 3rd. Series, 35: 110-44.

Barrow, J. (1801) *An Account of Travels into the Interior of Southern Africa*, London: T. Cadell, Jr. and W. Davies.

Boserup, E. (1965) *The Conditions of Agricultural Growth*, London: George Allen and Unwin Ltd.

de Schlippe, P. (1956) *Shifting Cultivation in Africa: The Zande System of Agriculture*, London: Routledge and Kegan Paul.

Giliomee, H.B. (1979) "The eastern frontier, 1770-1820," in Elphick, R. and Giliomee, H.B. (eds.), *The Shaping of South African Society, 1652-1820*, Cape Town: Longman-Penguin.

Giliomee, H. (1981) "Processes in development of the South African frontier" in Lamar and Thompson, *Frontier in History*, 76-119.

Gray, L.W. (1932) *History of Agriculture in the Southern United States to 1860*, Gloucester, Mass.: Peter Smith, 1958. Reprint of first edition 1932.

Guelke, L. (1976) "Frontier settlement in early Dutch South Africa," *Annals*, Association of American Geographers, 66: 40-41.

Guelke, L. (1985) "The making of two frontier communities: Cape Colony in the eighteenth century," *Historical Reflections/Réflexions Historiques*, 12: 434-48.

Lamar, H. and Thompson, L. (1981) *The Frontier in History: North America and Southern Africa Compared*, New Haven and London: Yale University Press.

Lefler, H.T. and Powell, W.S. (1973) *Colonial North Carolina* New York: Charles Scribner's Sons.

Lichtenstein, H. (1928-1929) *Travels in Southern Africa in the Years 1803, 1804, 1805 and 1806.* Translated by Plumptre, A., Cape Town: Van Riebeeck Society, 1928-1929.

Mitchell, R.D. (1977) *Commercialism and Frontier: Perspectives on the Early Shenandoah Valley*, Charlottesville: University Press of Virginia.

Neumark, S.D. (1957) *Economic Influences on the South Africa Frontier, 1651-1836*, Stanford: Stanford University Press.

Norton, W. and L. Guelke. (1977) "Frontier agriculture: subsistence or commercial?" *Annals*, Association of American Geographers, 67: 463-67.

Rice, O.K. (1970) *The Allegheny Frontier: West Virginia Beginnings, 1730-1830*, Lexington, Kentucky: The University Press of Kentucky.

Richards, A.I. (1939) *Land, Labour and Diet in Northern Rhodesia*, London: Oxford University Press.

Rohrbough, M.J. (1978) *The Trans-Appalachian Frontier: People, Societies and Institutions 1775-1850*, New York: Oxford University Press.

Sparrman, A. (1977) *A Voyage to the Cape of Good Hope, 1772-76*, Forbes, V.S. (ed.), Cape Town: Van Riebeeck Society.

Turner, F.J. (1961) "The significance of the frontier in American history," in Billington, R.A. (ed.), *Frontier and Section: Selected Essays of Frederick Jackson Turner*, Englewood Cliffs, New Jersey: Prentice Hall, Inc.

Wertenbaker, T.J. (1963) *The Old South*, New York: Cooper Square Publishers, Inc.

CHAPTER 12

FORCED HOUSING: A VIEW AT A PERENNIAL WORLD-WIDE MALIGNANCY

Ludwik Straszewicz
Department of Economic Geography and
Institute of Spatial Organisation
University of Łódz
and
University of Paris I
Pantheon - Sorbonne

Geographers, sociologists, economists, town planners and other scientists have analysed in depth the organisational problems related to the built environment. They attempt to discover laws governing human settlement and to formulate theories about places and networks; they discover that the location for urban centres depends on environmental factors; they discover the role of biophysical factors in town development, the coordination between a site for housing and the presence of large or small rivers and streams, of hills and valleys of plains and forests, they formulate principles on the dependence of urban networks on tertiary industries, highway, and railroad networks, and so on. Yet, all this research is primarily concerned with the free world with people who can choose freely where they want to live. It is a fact that all individuals are subject to economic forces: hunger and poverty direct their moves; ambition and feelings often play an important role in their choice of lifestyle. Yet, parallel to this world there is another world where power decides where people should live, even against their will. Within this ecumene, "forced" housing occupies a large area, far larger than we know about or even imagine.

Forced housing is a historical fact! We find its presence in the most remote times. The ancient Egyptian scripts, the great work of Homer, and the Bible give examples of it. During the Middle Ages, the new feudal Europe, born out of the barbaric invasions of the Roman Empire, saw this situation happening as it afforded the opportunity to build new states, new boundaries, a new social hierarchy and a new spatial structure. For centuries, peasants lost their independence while they were placed under the tyranny of the nobility who put them in bondage. They were denied the right to move freely, to leave their house or the farm where they worked, and where their fathers worked.[1] Peasants were attached to a specific place as ordered by the Lord.

Another example of forced housing was the Jewish ghetto. It was a well delimited district separated from others where the Christian population lived. The Jewish people were not allowed to live anywhere else. They could move freely as long as they were not leaving the ghetto. Both legal ghettos and de facto gettos inhabited by diverse ethnic groups, existed until World War II in many cities throughout Europe and United States. Since then, they have disappeared. Yet we can see something similar happening in many cities of Western Europe with the establishment of working class immigrants districts. While taking different shapes, these two types of forced housing can be seen throughout the centuries. There are also types which integrate both features. In one of these classes, we can find the case of the war prisoners who, once the hostilities ended, stayed in the victorious country. They were obliged to look after the land and the cattle and to plow the fields.

In other countries they had to be the new settlers of empty and uncivilised territories. The common justification behind these types of housing has been economic, political, and even ideological. People living in forced housing are people who are suspect, and by restricting their movements, it is easy to keep watch on them, to separate them from the rest of the population, or to control and force them to work.

Since the Middle Ages, we have seen freedom restrained in every way possible in almost every country, particularly in autocratic countries. In the ancient Ottoman Empire, the land was populated much the same way as during the Middle Ages when the peasants were assigned to a specific area. In the Balkan States, we can still find today these same marks left by Turkish political power. Russian history since its beginning, which is to say, upon the liberation of the Russian people from the Mongol yoke in the sixteenth century, shows a colonisation pattern of the country based on the method of forced housing which was used to dominate and control oppressed countries.

---

[1] They were only allowed to go to the city to try to find some shelter and there, after a period of one year and one day, they could be released from the feudal yoke and regain their freedom.

Forced housing was also established in the Prussian state by Frederic II and his successors.

Here the same pattern prevailed, although different forms were tried out for political reasons. There were areas forbidden to certain groups of people. For instance, Jews could not live in the centre of the Russian Empire. Since the beginning of the seventeenth century they were forbidden to settle in Spain. European history gives examples of Catholics and Protestants being expelled from their land. Both were denied the right to stay or settle in certain parts of the country whether it happened in France or Great Britain.

Contemporary history also gives us numerous examples of violation of human rights and of the freedom to choose one's living place. Germany, under the Nazi government of the Third Reich which was practicing nationalist politics, represents one of the most striking examples. Even before World War II, and following a signed agreement between the Italian and German governments, thousands of German people from South Tyrol were transferred from Italy to Germany. Theoretically, they were given the choice to settle where they chose, but in fact they had to settle in the areas planned in advance by the state organisation which administered the whole operation.

When the war broke out, the Hitlerian government made use of these previous experiences and displaced population movements occurred on a very large scale. This included an incredible migration of the population. Millions of men, women and children were forced to leave their houses, their farms, their apartments, and their country for a period of several years and to migrate. Some were sent away from the great German fatherland, others - the so-called "repatriates" - arrived from foreign countries. Even toward the end of 1939, thousands of Polish people were expelled from the western cities and villages which became part of the Reich. In these same places, they moved in German people from the Baltic States, who were forced to leave their resident country. It was the case, that they were not given the choice of the area in which they were going to live. Germans from Transylvania, Hungary and the Baltic States were following them. In return, millions of Germans were displaced from Polish and Czech territories.

Forced migrations were even more numerous in East Europe. Upon the execution of the Ribbentrop-Molotov agreement, the Soviets occupied half of Poland in the East, the Balkan States and later Bessarabia. Right at the start, thousands of people were arrested and deported to the East. They filled the camps or had to settle in the towns and villages in the far North of Europe, in Siberia and

Kazachstan. Most of them were Polish, yet others were with them too. Some of these unfortunate people, after long and cruel hours of transportation, years of hard work in the coal mines, the forests and the fields, were lucky enough to be freed and evacuated through Iran and the Far East to England, the United States, Canada, and other countries in America and Africa.

When the war ended, the Big Three agreed to change political boundaries which, in turn, caused massive population migrations. All this is well known, yet, it is necessary to recall the memory of it. The German minority who, before the war, lived in Poland and in Czechoslovakia, disappeared. German people, so called Volkdeutshce, were expelled: one million from Poland and three million from Czechoslovakia. Poland was condemned to give to the U.S.S.R., the Eastern part of its national territory, about 180,000 km². In compensation, 100,000 km² of pre-war German territory in the West was added to my country. This territory, which, in the past belonged to Poland and still bears the cultural relics of our ancestors, had the majority of its population composed of Germans who were forced to leave. Polish people replaced them either voluntarily or under compulsion. They came from East Poland which then became Russian.

Forty years have gone by since these events took place. The newcomers have become the old timers. Many of them are no longer living at the same place, yet they don't think that the problem is over. The problem of emigration of Poles from Silesia is still an issue between the Warsaw and Bonn governments. People are exchanged for money.

In Europe forced housing still exists! In the fourteenth century, the Tsarist government conquered Siberia through deportations. The deportees, mostly Polish political prisoners, worked in the mines. Those who survived these long periods of condemnation had to stay in special zones established for people suspected or disliked by the government. Most of them had to stay in Siberia. They populated villages and towns. Many got married and raised a family. Their descendants, although being "Russianised," testify today of their origins. Itkuck, for instance, similar to other cities, was built by the deportees. What deported people brought to the development of Siberia, is enormous. Many of them conducted research and the great discoveries made in Siberia during the nineteenth century are due to them. One can quote the pleiad of geologists, geographers, naturalists, biologists, ethnographers, etc. Names, like Czerski, Dybowski, Bronistaw, Pitsudski, Sieroszewski, as well as many others, are honoured in the Annals of the Imperial Academy of Science in

St. Petersburg. Mountains, cols, and islands bear the names of these unfortunate people (Straszewicz, 1973, 1975). Roger Brunnet has recently published an article on the role of prisoner camps in this country's economy. One may conjecture that the economic role played by the forced housing system is much larger.

The "forced" housing system also exists in the other hemisphere. Apartheid is a system based on racial separation: abolished a century ago in the United States it is still alive in the Republic of South Africa. We can even consider that the greatest percentage of the world population lives in forced housing. Forced housing is always closely connected to autocratic systems and occurs in countries dominated by authoritarian regimes. Yet it is also present in democratic countries. For instance, in France, the Basque people can only live in some of the southwestern regions, while, on another end, diverse groups of foreign people have a limited choice regarding their residential locations.

How is the forced housing system reflected in the social and economic environment? There is no clear answer as detailed data are not available. In the countries where it is occurring, it is carried out discreetly. There it takes shape. It is formed via police orders, yet it is a well known phenomenon and it deserves the interest of scientists, not only of the political scientist, sociologists and psychologists, but also of economists and geographers.

REFERENCE

Straszewicz L. (1975) "Les voyageurs et explorateurs polonais," Waclaw Slabcznski: Polscy podroznicy i odkrywcy, Warszawa 1973, PWN, 465 pp. Compte-rendu: *Annals de Geographie*, Nr. 462, LXXXIV: 239-40.

PART IV

URBAN SYSTEMS

CHAPTER 13

POSSIBLE CONTRIBUTIONS OF A COMBINED ECONOMIC BASE-CENTRAL PLACE THEORY TO THE STUDY OF URBAN SYSTEMS [1]

Richard E. Preston
and
Clare J.A. Mitchell
Department of Geography
University of Waterloo

This study continues to explore an approach to the study of urban systems begun by Illeris and Dziewonski during the 1960s. They suggested that base theory and central place theory could be combined into a single framework for studying the development of urban systems. During the 1970s, Dziewonski and Jerczynski considered this possibility on empirical and theoretical grounds and continued to endorse it as a partial answer to the need for a general theory of urban systems.

Illeris (1964) and Jerczynski (1973, p. 55) identified several instances of logical consistency between concepts fundamental to both base and central place theories; e.g. between Christaller's concept of "Centrality," or central place importance, and the key concept of "Basic," or export activities, in base theory. Moreover, using a single conceptual framework, Illeris applied both theories in a study of "The Functions of Danish Towns." Dziewonski and Jerczynski built on Illeris's research and identified advantages of a combined base and central place theory for urban systems research: (1) base theory, which dealt originally with individual settlements, could be widened to

----

[1] Based on a paper presented during the Second Meeting of the I.G.U. Commission on Urban Systems in Transition, Pamplona, Spain, August 26-30, 1986.

embrace entire settlement systems, (2) central place theory, with its narrower focus, could be integrated into an overall urban systems context, and (3) the explanatory power of both theories could be enriched by complementing their traditional export/local explanation of activity importance by an additional subdivision of activities into standard and specialised types as suggested by Illeris (Dziewonski, 1967, pp. 144-45, 1978, p. 46, Dziewonski and Jerczynski, 1978).

Other scholars also commented on relations between the theories. Isard (1960, pp. 222-27) noted the failure in base studies to consider study areas as part of regional urban systems, and he suggested that one solution to this problem lay in developing relations between base and central place theories. Parr (1973, pp. 185-87) considered expansion of central place theory to include a larger share of economic activities. Preston (1977, 1986b, 1987), like Illeris (1964), Jerczynski (1973) and Carol (1960), noted the equivalence of Christaller's (1966, pp. 17-19) concepts of centrality and local consumption and the export and nonbasic components of base theory and suggested compatible measures. Borchert (1967) observed that applications of base theory were possible within a central place framework. Only Norse (1978), however, considered explicitly the question of the theories' equivalence. In this regard, it is surprising that no English language urban geography or urban economics textbook considers the relation between these theories.

Because existing research is exploratory, the possibility of a combined theory and its possible contribution to the study of urban systems represent open problems in the study of urban systems. Moreover, it appears that the chief task is to clarify the relationship between the theories. An attempt is made in this study, therefore, (1) to continue development of the foundation for combining base and central place theories laid by Illeris, Dziewonski and Jerczynski, (2) to summarise other pertinent research, and (3) to identify critical research foci. These tasks are complicated because base and central place theories are not equivalent either in the range of activities embraced by their domains or in their treatment of normative settlement patterns.

This study proceeded from the assumption that a basis for combining base and central place theories exists in their "Partial Equivalence." For example, their explanations of the population size and functional importance of nodes, on the one hand, and of the formation and structure of urban systems, on the other, appear to be equivalent. Moreover, there are clear relations between the goals and logical structure of both theories and key research areas in the urban systems approach; for example, the definition of spatial components, boundaries, the role of external relations, the structure of interdependence, processes of growth and change, and form and impact of public intervention and control (Simmons and Bourne, 1981, p. 421).

Given partial equivalence between the theories and their shared tasks in the systems approach, several broad questions were examined: (1) To what extent can base theory serve as a holistic explanation of the functional importance and spatial interdependence of nodes in an urban system? (2) Can central place theory serve as the most general component of such an overall explanation? And (3), within such a holistic framework, can the general locational structure of urban systems be accounted for by an additative application of partial location theories along lines suggested by Christaller (1966, p. 7) and Dziewonski (1967, 1970, 1978)?

BASE AND CENTRAL PLACE THEORIES COMPARED

There are three fundamental differences between base and central place theories. First, base theory is a holistic theory of city size while central place theory is a partial theory. Second, base theory has been perceived as essentially aspatial while central place theory has a well defined spatial dimension. Third, only central place theory has an abstract-normative component that generates "Ideal Type" settlement patterns.

Accordingly, base theory accounts for any urban activity, regardless of how standard or sporadic its occurrence may be. Central place theory, by contrast, accounts for only standard activities that serve extra local as well as local consumers, that agglomerate regularly in settlements, locate according to the "nearest centre" behavioural assumption, and occur in level representative clusters in central place hierarchies in response to variations in threshold size. Consequently, central place theory was intended to account for the location of retail and service trades, wholesaling, transportation and communication activities, public and cultural activities, some commercial and financial activities, and certain types of market oriented industries (Christaller, 1966, pp. 7, 20-21, 140-41).

The goal and structure of base theory have remained essentially unchanged since its introduction in this century by Sombart in 1907 (Dziewonski, 1967, Borchert, 1967). By contrast, numerous theories of central places have been developed since the appearance of Christaller's basic work in 1933 (e.g. Lösch, 1954, Berry and Garrison, 1958, Beckmann, 1958, Berry, 1967, Beavon, 1977, Smith,

1965, White, 1974, Alan and Sanglier, 1979), and they diverge in their claimed explanatory powers, which range from Christaller's specifically defined regional importance (centrality) to overall nodal size. It is noteworthy, however, that the areas of overlap, or partial equivalence, between the theories considered here exist explicitly only in Christaller's theory. Hence, references to central place theory refer to Christaller, unless stated otherwise.

By contrast with their differences, base and central central place theories have at least five fundamental areas of overlap, which comprise their state of partial equivalence: (1) they were intended to explain settlement importance at the intercity scale, (2) they cover central place activities, (3) their explanations of nodal importance are logically consistent, (4) they focus on interurban flows which, in turn, identify urban systems, and (5) they offer both static and dynamic systems perspectives. Because it covers a common but more specific set of urban activities, therefore, central place theory can be conceptualised as occupying a subdivision of base theory.

The theories describe and explain functional importance in a consistent manner. Activity or nodal importance (size) is attributed to two sources, that part accounted for by local demand and that part accounted for by external demand. Status or change in functional importance or size can be described and explained in both theories by these components and their dynamic relationship. Moreover, the roles of both local or external forces in nodal stability and change can be monitored and emphasised. These points are demonstrated by the models in Table 1.

Both theories emphasise and characterise nodal importance in terms of "Surplus Functional Importance." In base theory, surplus activity is designated basic or export, while in central place theory it is called centrality. In both theories surplus activity represents a flow of goods and services from a centre to consumers not located in that place, but which are located within a larger urban system defined by such flows.

While both theories explain the importance of activities and nodes by magnitudes of surplus goods and services, they also show the contribution of nonbasic or local demand to the total population size and functional importance of a place. Thus, their common two-part explanation of nodal size or welfare can be used in the traditional base theory framework where explanatory power is attributed to the surplus component, or in a more flexible framework where causality can be attributed to either basic or nonbasic components or to any combination thereof (Tiebout, 1962). The hypothesis underlying base theory is simply changed from, urban size is a function of basic

Table 1

Comparison of the Structure of Operational Base and Central Place Models

In Economic Base Theory	In Central Place Theory
(1) TA = BA + NBA	N = C + L
(2) BA = TA - NBA	C = N - L
(3) BTR = BA/TA	CNR = C/N
(4) BNR = BA/NBA	CLR = C/L
(5) $BAM = \dfrac{1}{1 - \dfrac{NBA}{TA}}$	$CPM = \dfrac{1}{1 - \dfrac{L}{N}}$
where:	where:
TA = total activity in a place (NODALITY)	N = total central activity in a place (NODALITY)
BA = surplus (export) activity in a place (BASIC)	C = surplus (export) central activity in a place (CENTRALITY)
NBA = locally consumed activity in a place (NONBASIC)	L = locally consumed central activity in a place (LOCAL CONSUMPTION)
BTR = basic/total activity ratio	CNR = centrality/nodality ratio
BNR = basic/nonbasic ratio	CLR = centrality/local consumption ratio
BAM = economic base multiplier	CPM = central place multiplier

activity, to urban size is a function of some combination of basic and nonbasic activity.

Both theories are dynamic and can thus serve as bases for treating settlements as components in evolving hierarchical and/or nonhierarchical settlement systems. Therein, groups of places are organised around system-forming activities or nodes that are linked and organised spatially by flows of surplus activity.

RESEARCH ON RELATIONS BETWEEN BASE AND CENTRAL PLACE THEORIES

Only Norse (1978) has studied specifically the "Equivalence of Central Place and Economic Base Theories of Urban Growth." He proceeded from the finding that "Curiously, no one has shown how the two theories relate to each other" (Norse, 1978, p. 543). He concluded (1) that the theories were equivalent, (2) that because of this, previous criticisms of economic base theory also apply to central place theory, and (3) that accordingly, both should be considered discredited theories of urban growth.

It is suggested that Norse's study, and thus his criticism, was flawed from the outset by adoption of two false assumptions regarding central place theory: first, that central place, like base theory, is a theory of overall city size, and, second, that the magnitude of a city's population is a reasonable measure of central place importance (Norse, 1978, p. 543). Both assumptions are inconsistent with the definitions and logic of Christaller's theory. This is a fundamental point because the same questionable assumptions characterise other studies in which relations between the two theories are mentioned, and, more importantly, because their continued adoption masks key areas of "partial equivalence" between the theories that are absent from reformulations of Christaller's original statement. Norse did not cite Christaller.

Besides Norse's specific inquiry, three research streams involving relations between base and central place theories have appeared since 1958.

Base and Central Place Theories and Urban Functions

In this research, a combination of base and central place theories is viewed as desirable and achievable. Isard (1960, pp. 222-27) noted that base studies failed to look outside of cities and regions under study and to consider them as part of existing urban and regional hierarchies. He suggested that a solution to this deficiency lay in development of relations between base theory and the nested hierarchy of settlements called for by central place theory. Illeris (1964) recognised the conceptual equivalence of key concepts in both theories, used the minimum requirements procedure to identify basic and nonbasic activity, on the one hand, and central place importance, on the other, and took this approach a step further by characterising basic and nonbasic activities as either "Standard" (ubiquitous) or "Specialised" (sporadic).

Dziewonski and Jerczynski (Dziewonski, 1967, 1970, 1975, 1978, Dziewonski and Jerczynski, 1978, Jerczynski, 1972, 1973) recognised explicitly that base theory is a holistic, and central place a partial, theory of nodal size. Moreover, they took into account the structural similarities and differences between the theories noted in Table 1. They combined the theories by hypothesising and demonstrating that nodal interdependence within national and regional urban systems could be explained by the hierarchical structure of central functions and by the territorial division of labour throughout a nation expressed by specialised urban functions. They suggested that such an approach provides a reasonably realistic theory of post-industrial settlement patterns, one that accounts, first, for patterns of functional and spatial organisation that are evolving toward more integrated states, and, second, for the specialised activities that are of increasing importance in the development of new settlement patterns. Their research also revealed, like Illeris (1964) and Pred (1975, 1976, 1977), that location patterns of numerous specialised functions do not conform with hierarchical structures suggested by central place theories.

Dziewonski and Jerczynski, once again building on Illeris, also concluded that by equating "Standard" activities with local and regional activities and "Specialised" activities with supraregional roles, a framework was created for integrating base theory, to account for specialised functions, with central place theories, to account for standard functions. This union, they suggested, would provide a foundation for developing dynamic models of urban system stability and change. In their approach, they assumed (1) that settlement systems form open economic regions in which the distinctive roles of exogenous and endogenous activities are decisive, (2) that the explanatory power of base theory is enhanced by inclusion of the concepts of standard and specialised activities and by determination of their local, regional, and supraregional roles, and (3) that a base theory umbrella could account for development of centres based on both standard and specialised functions.

Measurement of Central Place Importance

Carol (1960) in a study of the business hierarchy of Zurich, Illeris (1964) in a study of the function of Danish towns, and Preston (1977, 1986b, 1987), as a byproduct of studies of central place patterns defined according to Christaller, noted that: (1) Central place theory's emphasis on centrality and links between towns and their complementary regions placed it under the base theory approach, (2) this arrangement emphasised base theory's ability to identify nodal interdependence, (3) the part of a town's total production of central goods and services accounted for by centrality could be considered basic and the part accounted for by local consumption could be considered nonbasic, (4) the centrality/nodality ratio provides an indication of the central place function's contribution to a nodal economy, and (5) given the dynamic and normative dimensions of central place theory, its combination with base theory should enhance a united theory's ability to explain nodal size and importance and urban system stability and change.

Explanation of City Size in an Urban Hierarchy

"City Hierarchies and the Distribution of City Size" has been a fertile research area since Beckmann's publication under that title in 1958. Although focussing on other matters, much of this research dealt indirectly with relations between base and central place theories. Beckmann (1958, p. 243) examined the hypothesis that "the size (population) of any city is proportional to the population it serves (including that of the city itself)," and he demonstrated that an urban multiplier could be used to determine the population of cities at different hierarchical levels. Despite emphasis on urban hierarchies and similarities between his model and the economic base model, Beckmann did not mention base theory, the question of equivalence between base and central place theories, or Christaller.

Beckmann and McPherson (1970) continued research on the distribution of city size in urban hierarchies. They defined the hierarchy under study as a central place hierarchy, and although noting some of Christaller's reservations regarding the use of population as a measure of central place importance, continued to focus on it. Neither central place importance according to Christaller nor equivalence between base and central place theories were considered explicitly. However, Beckmann and McPherson did suggest that a useful analytic extension of their model would be relaxation of the assumption that centres operate only to serve themselves and their immediately surrounding hinterland.

Dacey (1966, p. 27) noted that Beckmann's (1958) study of the distribution of city size in an urban hierarchy was of particular interest because it offered the only model which based city size on the population of it's market area. Dacey (1966, 1970, Dacey and Huff, 1971) undertook to explain the population size of places at particular levels of a precisely defined hierarchy. Dacey's (1970) approach was to "construct models that related central place structure and economic base theory." Dacey meant the construction of models explaining urban population size based on an integration of alternative formulations of economic base theory with the geometric structure of a hierarchical central place system of the Christaller type. Dacey demonstrated that population size at different levels in such a central place system could be explained in base theory terms. He did not, however, attempt to explain either the central place importance of nodes in a central place hierarchy or the relationship between base and central place theories.

In the early 1970s, this research split along three overlapping paths. One continued the search for an explanation of both the population size and overall functional complexity of places in urban (often central place) hierarchies, and solutions were sought and expressed increasingly in base theory terms. A related path also strove to explain city population size in urban hierarchies. However, it differed fundamentally from the first stream in that, like Norse (1978), its participants assumed that "Theoretical conformance typifies the economic base and central place models" (e.g. Horn and Prescott, 1978, p. 230, Thompson, 1982). Accordingly, they assumed that both theories provided explanations of overall city size and that urban population size and central place importance were equivalent. They, thus, failed to recognise, as researchers in the Beckmann-Dacey stream apparently did recognise, that, unlike base theory, central place theory was not a holistic theory of city size.

A third path emphasised the estimation of multipliers. Beckmann (1968), Dacey (1966), Beckmann-McPherson (1970), Dacey and Huff (1971), and others estimated city populations by applying what they called population and service multipliers. In some cases, multipliers were constant among hierarchical levels and in some cases variable. Estimation of multipliers generally, for different hierarchical levels, for different levels of sectoral disaggregation, and for different technological conditions, generated a flurry of studies (Mulligan, 1984, Richardson, 1985).

Key points regarding the base-central place theory dimension of research on the distribution of city population size in urban hierarchies are: (1) models were built to derive city populations in urban hierarchies (often specified as central place hierarchies), (2) variations of base theory were used to explain and estimate city population size and functional complexity for centres at different hierarchical levels and under different conditions, but, (3) attempts were not made to explain central place importance in Christaller's terms or to consider equivalence between the theories. This interpretation appears to be shared by Parr, Denike, and Mulligan (1975) and Mulligan (1984, p. 19).

STEPS TOWARD A COMBINED THEORY

A combined base-central place theory covers all activities in any urban system and is capable of producing specific answers for particular activities, nodes, or subsystems (Figure 1). While this framework has substantial explanatory power and is flexible, it is not without deficiencies, three of which are: (1) a lack of total equivalence between the theories' domains, (2) a combined theory's inability to generate normative territorial patterns for all urban activities, and (3) the extent to which criticisms of central place and base theory compromise a combined theory. The first two of these are considered below.

Lack of Total Equivalence Between the Theories' Domains

One way to reduce the partial equivalence problem is to expand the domain of central place theory. This is achieved by incorporation of previously excluded and unemphasised activities, an approach that also holds promise for expanding the normative dimension of a combined base-central place theory.

Figure 1: A Base Theory-Location Theory Framework for Activities in Urban Systems

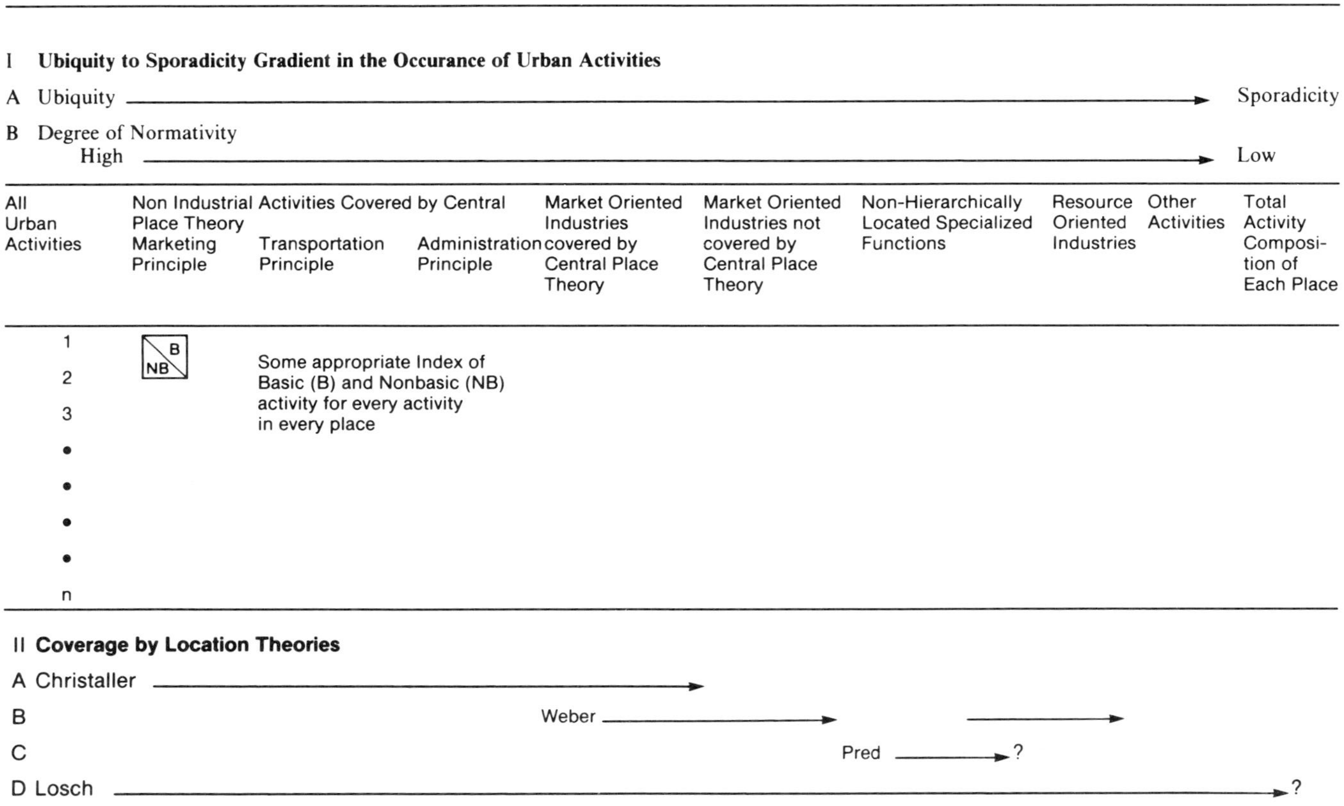

I Ubiquity to Sporadicity Gradient in the Occurance of Urban Activities

A Ubiquity ——→ Sporadicity

B Degree of Normativity
High ——→ Low

All Urban Activities	Non Industrial Place Theory Marketing Principle	Activities Covered by Central Transportation Principle	Administration Principle	Market Oriented Industries covered by Central Place Theory	Market Oriented Industries not covered by Central Place Theory	Non-Hierarchically Located Specialized Functions	Resource Oriented Industries	Other Activities	Total Activity Composition of Each Place
1	B/NB								
2		Some appropriate Index of Basic (B) and Nonbasic (NB) activity for every activity in every place							
3									
•									
•									
•									
•									
n									

II Coverage by Location Theories

A Christaller ——→

B Weber ——→ ——→

C Pred ——→ ?

D Losch ——→ ?

One approach is to consider candidates from classes of activities with location patterns divergent from those generated by central place theory, for example, manufacturing. Given the increasing importance of the market as a factor in the location of manufacturing plants, it is suggested that there are grounds for incorporation of at least some of the increasing number of market-oriented industries under the domain of central place theory.

As noted, Christaller's theory was intended to cover a specific range of urban functions, and thereby supplement Weber and Thunen's theories of manufacturing and agricultural location in any explanation of the structure of an overall space economy. However, Christaller (1966, pp. 7, 20-21, 46, 71, 96-97) both explained why the location of some types of manufacturing belonged under his theory, and discussed the pertinence of his theory to industrial location in general. On the first point, he stated:

> Above all, those manufacturers which, according to Alfred Weber's terminology, are consumer-oriented (excluding those which can be found in almost any village), belong in central places; . . . Centre-oriented production occurs when the collection and further manufacture of dispersedly produced unfinished goods or raw materials occur at the centre. Typical illustrations of centre-oriented production are the many handicraft industries, frequently the food industries, breweries, dairies, sugar refineries, and container industries. Breweries are consumer-oriented and dairies, sugar refineries, and container industries are raw material-oriented. (Christaller, 1966, p. 20)

Christaller (1966, pp. 140-41) also mentioned newspapers, mills, bakeries, slaughter houses, gas works, and electric (steam) plants as possible market-oriented industries.

Christaller's observations provide a base for expanding his theory's domain. By identifying types of manufacturing that he thought should fall under his theory as market-oriented, on the one hand, and raw-material oriented, on the other, he invited attempts to both generate a list and to monitor types of manufacturing that locate according to central place principles. This problem has been studied primarily in relation to Christaller's market principle. For example, Faust (1975) based his examination of "Ubiquitous Manufacturing" on criteria of frequency-of-occurrence and market-orientation. He produced a list of ubiquitous (market-oriented) manufacturing activities that, in his view, "should characterise the lowest level of a hierarchy of market-oriented manufacturing" (Table 2).

Table 2

Market-Oriented Industries and Their Major Locational Controls

Locational Controls

1	2	3	4	Industry
X	X	X		Ice cream and frozen desserts
X	X	X		Fluid milk
X			X	Prepared animal feeds
X	X			Bread and related products
X	X	X		Bottled and canned soft drinks
X				Millwork plants
	X	X	X	Newspapers
			X	Printing, except lithographic
			X	Printing, lithographic
X		X		Concrete block and brick
X		X	X	Other concrete products
X	X	X	X	Ready mixed concrete
X			X	Fabricated structural steel
X			X	Sheet metal work

Locational Controls

(1) Manufacturers of finished products which are much more expensive to transport than the raw materials from which they are made;

(2) Manufacturers of goods which are significantly greater in weight, fragility, or perishability than the raw materials from which they are made;

(3) Manufacturers where the raw materials used in the manufacturing process are relatively ubiquitous with respect to population; and

(4) Manufacturers of goods which require close consultation between manufacturer and consumer in terms of size, quantity, and/or design. (Source: Faust, 1975, 16).

Faust (1975, p. 17) suggested that these industries occupy numerous specific locations that mirror the distribution of population. He noted, as did Christaller, that "Both the sizes and locational frequency of these markets result from the interaction between transportation costs and scale economies as well as areal variations in demand (population density, incomes, and tastes)." Faust's list of market-oriented manufacturing activities thus falls under Christaller's market principle and expands his suggested list. Faust's approach further enriches a combined base-central place theory by suggesting locational controls elaborating Christaller's nearest centre-threshold control (Table 2).

Christaller also suggested that at least some types of raw material-oriented manufacturing (dairies, sugar refining) should fall under his theory. Because Faust was interested primarily in what he defined as ubiquitous manufacturing, he rejected the raw material-oriented class because (1) their national location pattern did not mirror the distribution of population, (2) they did not utilise the same specific raw materials, or (3) produce comparable finished products.

Nevertheless, Faust's study revealed that even though the distribution of some types of raw material-oriented plants did not correspond to the overall pattern of population, both high frequency and spatial regularity of occurrence could result if a critical input was found in a large number of locations in particular regions. Examples of such manufacturing activities that were eliminated in Faust's study were producers of forest products (logging camps and contractors, saw mills, and planing mills), fruit canning, and vegetable processing plants. Both forest products and food processing reveal patterns of regional specialisation, and may conform to central place theory within those regions.

Parr (1973, p. 186) considered both how Christaller's theory accommodated certain market and raw material-oriented industries and suggested that the theory could be modified to embrace more fully activities whose locations are influenced by access to inputs and supply area considerations. Parr suggested that such a broadening of the scope of central place theory would also permit it to be used to analyse locational patterns of activities which contribute to a centre's economic base less regularly than pure market-oriented industries. Such modification offered scope for bringing regionally specialised industries at least partially under the domain of central place theory. Parr's examples included agricultural industries (flour milling, meat packing, cotton ginning, and dairy production) and forest product industries. Parr noted that Beckmann (1968, Chapter 5) also considered modifications of central place theory along these lines.

Faust's results, and additions thereto from research by Christaller and Parr, designate mainly the more common and thus lower order forms of market or raw material-oriented industries. Faust (1975, p. 17) suggested that discovery of a complete hierarchy of market-oriented (we would add, and region specific raw material-oriented) industries would be a useful research direction. If such a hierarchy exists, this research could expand the domain of central place theory and thus expand the degree of functional and normative equivalence between base and central place theories.

The Industrial Hierarchy Problem

Research both confirming Faust's findings and exploring the nature of a possible industrial hierarchy has begun. As part of a larger study, Koenig, Lewis, and Ray (1975, p. 182) attempted to define a manufacturing hierarchy for Canadian cities of over 10,000 population in 1961. All industries at the three digit Standard Industrial Classification (SIC) level were ranked by frequency of occurrence criteria and divided into five levels (fully ubiquitous, and four equal-range ubiquity classes). Cities were allocated to the five-level hierarchy according to "minimum incidence-requirements" set for each level for each set of industries. They found clear relations between industrial incidence and their arbitrary hierarchical levels and were able to classify activity groups along a range from ubiquitous to sporadic. Scorrar and Williams (1976) examined relations between manufacturing activity and the urban hierarchy. They based their analysis of Canadian cities on measures of manufacturing threshold, concentration, and incidence, and they concluded:

> That there is a strong relationship between the type and number of industries in an area and the population of that area. In general, the smaller towns have only ubiquitous industries. The larger the centre, the greater the likelihood of many industrial types, including some sporadic industries being present. (Scorrar and Williams, 1976, p. 112)

Their findings regarding industrial occurrence by size of urban area, along with the more general results of Koenig, Lewis, and Ray, support the suggestion that research on the nature of an industrial hierarchy could contribute to expansion of the domain of central place theory and, thereby, to greater equivalence between base and central place theories.

SUGGESTED RESEARCH FOCI

Numerous possibilities exist for research leading to greater equivalence between base and central place theories through an expansion of the latter's domain. For example, there is much to be learned about both the locational behaviour and basic (export) component of the so-called post-industrial activities (Naisbitt, 1982, pp. 12-13), the expanding service industries (Daniels, 1985, Illeris, 1986), cultural activities (Mitchell, 1987), quaternary activities in general, and about high order business and financial activities in particular (Semple and Green, 1983, Preston, 1986).

It is also surprising that there is not a comprehensive treatment of activities that locate according to Christaller's transportation principle. Knowledge of this class of activity (e.g. wholesaling, storage, warehousing, break-of-bulk based processing industries, transportation terminals and equipment repair activities) could also expand central place theory's domain, equivalence between the theories, and the normative capacity of a combined theory.

Despite attempts by Christaller to forge links between his theory and the location of manufacturing, it was clear from the outset that central place theory was not intended to account for all forms of industrial location. Christaller's (1966, p. 69) awareness of this situation was reflected in his observation that in the interpretation of space economies, factors causing change and variation in, and deviation from, his central place system must be studied, and that one such key factor is manufacturing. This is still the case. Despite valuable findings by the studies reported here, relations between central place patterns, on the one hand, and primary manufacturing patterns, secondary manufacturing patterns, and other forms of nonhierarchically organised activity, on the other, are poorly understood.

Reinforcing the observations by Illeris, Dziewonski and Jerczynski noted earlier regarding the nonhierarchical location of some specialised functions, Pred (1975, p. 124) suggested that in western urban systems, the spatial structure of multilocational organisations involves complex intraorganisational linkages between main offices and subordinate "domestic units" specialising in a particular aspect of production. Citing evidence from the United States, Britain, Germany, and Sweden, Pred suggested that the intraorganisational linkages of multilocational corporations are not symmetrical; i.e. they do not necessarily extend from a major metropolitan centre through the hierarchy to successively lower order centres. Rather, these linkages are asymmetrical, extending between several metropolitan centres or to smaller centres beyond the immediate hinterland. Pred argued that because central place theory can account only for hierarchical organisation and diffusion (symmetrical diffusion down the hierarchy), it is inappropriate for explaining the complex location of specialised functions of the multilocational firm. Given Christaller's clear recognition of the importance of nonhierarchically organised activities, it is not surprising that their number may be increasing along with the complexity of social and economic organisation in general and that theory is needed to explain their location patterns.

Thus, the nature of relationships between, and the possible additivity of, various forms of vertical spatial organisation represents an inviting research area. Fortunately, a few innovative studies of this problem have been completed and they have broken new ground. Starting with Christaller (1966), numerous scholars have recognised that the overall vertical organisation of settlements in space economies reflects the simultaneous operation of multiple hierarchical and nonhierarchical distributions of activities and places (Losch, 1954, Berry, Barnum, and Tennant, 1962, pp. 102-03, Woldenburg, 1968, Pred, 1975, 1976, 1977, Dziewonski, 1975, pp. 148-49, Parr, 1978). As a result, there have been attempts to develop alternative central place systems that modify and incorporate properties from the theories of Christaller, Losch and other central place scholars (Mulligan, 1984, pp. 11-12). Research on the development of such mixed settlement hierarchies is fundamental to development of a theory of urban systems.

Another research path involves the possibility of integrating a combined base-central place theory into an overall, and spatially explicit, theory of regional development. The suggested umbrella is Perroux's (1955) original theory of uneven regional development, or growth poles. This possibility has been explored by Parr (1973) and Preston (1984).

CONCLUSIONS

Possible bases for a combined economic base-central place theory, its possible contributions to the study of urban systems, and related research findings and opportunities have been considered. Support is offered for the position that base theory can serve as one overall framework for identifying and explaining urban systems. It is also suggested that central place theory can serve as the most general component of such a framework, and that its inclusion and domain expansion could strengthen the normative dimension of a combined theory. Finally, at least on conceptual grounds, there appears to be no reason why the spatial structure of activity in an overall base theory framework could not be explained, first, by an additive application of the marketing, administrative, transportation, and possibly mixed hierarchy principles from central place theory, and why nonconforming activities could not, then, be accounted for by a further additive application of other location theories (e.g. Weber and Pred) as suggested by Christaller (1966, p. 7) and Dziewonski (1967, 1970, 1978) and outlined in Figure 1. Such an overall framework should not only enhance research in economic and urban geography, it should facilitate teaching those subjects as well.

Finally, relations between a combined base-central place theory and the systems approach have been emphasised. This is significant because it appears that the systems approach provides an appropriate and robust framework with virtually unlimited potential for the study of urban systems. Mulligan (1984, p. 9), noted, for example, that a combined theory, with its emphasis on flows and nodal interdependence, its two-part explanatory framework, and its dynamic systems perspective, facilitates its integration with input-output theory, a union which places the new framework on "secure theoretical grounds." A combined base-central place theory, may also represent one route toward operationalising the n x n matrix describing interdependence among urban and regional systems that was suggested by Simmons and Bourne (1981, p. 421) as an appropriate framework for the analysis of urban and regional systems.

REFERENCES

Allen, P.M. and Sanglier, M. (1979) "A dynamic model of growth in a central place system," *Geographical Analysis*, II: 256-72.

Beckmann, M. (1958) "City hierarchies and the distribution of city size," *Economic Development and Cultural Change*, 6: 243-48.

Beckmann, M. (1968) *Location Theory*, New York: Random House.

Beckmann, M. and McPherson, J.C. (1970) "City size distribution in a central place hierarchy: an alternative approach," *Journal of Regional Science*, 10: 25-33.

Beavon, K.S.O. (1977) *Central Place Theory: A Reinterpretation*, London: Longman.

Berry, B.J.L. (1967) *Geography of Market Centres and Retail Distribution*, Englewood Cliffs: Prentice Hall.

Berry, B.J.L. and Garrison, W.L. (1958) "Recent developments of central place theory," *Papers and Proceedings of the Regional Science Association*, 4: 107-20.

Berry, B.J.L., Barnum, H.G. and Tennant, R.J. (1962) "Retail location and consumer behaviour," *Regional Science Association Papers*, 9: 65-106.

Borchert, J.G. (1967) "Basic and service: limitations of the concept and alternative developments," *Bulletin, Geografisch Instituut, Utrecht*, 4: 41-69 (translated by Malcom Matthew).

Carol, H. (1960) "The hierarchy of central functions within the city," *Annals*: Association of American Geographers, 50: 419-38.

Christaller, W. (1933) *Central Places in Southern Germany: An Economic Geographical Investigation of the Regularity of the Distribution and Development of Settlements with Urban Functions*, Jena: Fischer. Partial translation by Baskin, C.W. (1966), *Central Places in Southern Germany*, Englewood Cliffs: Prentice Hall.

Dacey, M.F. (1966) "Population of places in a central place hierarchy," *Journal of Regional Science*, 6: 27-33.

Dacey, M.F. (1970) "Alternative formulations of central place populations," *Tijdschrift Voor Econ. En Soc. Geografie* , 61: 10-15.

Dacey, M.F. and Huff, J.O. (1971) "Some properties of population for hierarchical central place models," *Tijdschrift Voor Econ. En Soc. Geografie*, 62: 351-55.

Daniels, P.W. (1985) *Service Industries: A Geographical Appraisal*, London: Methuen.

Dziewonski, K. (1967) "The concept of the urban economic base: overlooked aspects," *Regional Science Association Papers*, 21: 139-45.

Dziewonski, K. (1970) "Specialisation and urban systems," *Papers of the Regional Science Association*, 24: 39-45.

Dziewonski, K. (1975) "The role and significance of statistical distributions in studies of settlement systems," *Papers of the Regional Science Association*, 34: 145-55.

Dziewonski, K. (1978) "Analysis of settlement systems: the state of the art," *Papers of the Regional Science Association*, 40: 39-49.

Dziewonski, K. and Jerczynski, M. (1978) "Theory, methods of analysis and historical development of settlement systems," *Geographia Polonica*, 39; 201-09.

Faust, J.B. (1975) "Ubiquitous Manufacturing," *Annals*: Association of American Geographers, 65: 13-17.

Horn, R.J. and Prescott, J.R. (1978) "Central place models and the economic base," *Journal of Regional Science*, 18: 229-41.

Illeris, S. (1964) "The functions of Danish towns," *Geografisk Tijdschrift*, 63: 203-36.

Illeris, S. (1986) "How to analyse the role of services in regional development," in Borchert, J.G., Bourne, L.S. and Sinclair, R. (eds.), *Urban Systems in Transition: Netherlands Geographical Studies*, 16:45-59.

Isard, W., (1960) *Methods of Regional Analysis*, New York: Wiley.

Jerczynski, M. (1972) "The role of functional specialisation of cities in the formation of a settlement network," *Geographia Polonica*, 24: 31-44.

Jerczynski, M. (1973) "Problems of specialisation of the urban economic base of major cities in Poland," 9-134, *Prace Geogreficzne*, 97: (English Summary) 131-34.

Koenig, E.F., Lewis, J.S. and Ray, D.M. (1975) "Allometry and manufacturing hierarchy: a general system theory approach to manufacturing employment and industry incidence," in Collins, L. and Walker, D.F. (eds.), *Locational Dynamics of Manufacturing Activity*, New York: Wiley, pp. 159-99.

Lösch, A. (1954) *The Economics of Location*, New Haven: Yale University Press.

Mitchell, C.J.A. (1987) *An Exploration of the Relationship Between Urban Systems Evolution and Cultural Aspects* of Community Support, Waterloo: University of Waterloo, Department of Geography, unpublished Ph.D. thesis.

Mulligan, G.F. (1984) "Agglomeration and central place theory: a review of the literature," *International Regional Science Review*, 9: 1-42.

Naisbitt, J. (1982) *Megatrends: Ten New Directions Transforming Our Lives*, New York: Warner Books.

Norse, H.O. (1978) "Equivalence of central place and economic base theories of urban growth," *Journal of Urban Economics*, 5: 543-49.

Parr, J.B. (1973) "Growth poles, regional development and central place theory," *Papers of the Regional Science Association*, 31: 173-212.

Parr, J.B., Denike, K.G. and Mulligan, G. (1975) "City-size models and the economic base: a recent controversy," *Journal of Regional Science*, 15: 1-8.

Parr, J.B. (1978) "Models of the central place system: a more general approach," *Urban Studies*, 15: 35-49.

Perroux, F. (1955) "Note on the concept of growth poles," Trans. by Gates, L. and McDermott, A.M. in Mckee, D.L., Dean, R.D. and Leahey, W.H. (eds.), (1970), *Regional Economics: Theory and Practice*, New York: The Free Press, pp. 93-103.

Pred, A.R. (1975) "On the spatial structure of organisations and the complexity of metropolitan interdependence," *Papers of the Regional Science Association*, 35: 115-42.

Pred, A.R. (1976) "The Interurban transmission of growth in advanced economies: empirical findings versus regional-planning assumptions," *Regional Studies*, 10: 151-71.

Pred, A.R. (1977) *City-systems in Advanced Economies*, London: Hutchinson.

Preston, R.E. (1977) "Some measurement problems in empirical central place studies," in Eidt, R.C., Singh, K.N. and Singh, R.P.B. (eds.), *Man, Settlement and Culture*: Essays in the Honor of Professor R.L. Singh, New Delhi: Kalyani, pp. 248-65.

Preston, R.E. (1984) "Relationships between classical central place theory and growth centre based regional development strategies," in Bryant, C.R. (ed.), *Waterloo Lectures in Geography: Regional Economic Development*, Waterloo: University of Waterloo, Department of Geography Publication Series No. 23, pp. 73-115.

Preston, R.E. (1986a) "A business press perspective on the emergence of Toronto as Canada's dominant city," in Borchert, J.G., Bourne, L.S. and Sinclair, R. (eds.) *Urban Systems in Transition: Netherlands Geographical Studies*, 16: 70-84.

Preston, R.E. (1986b) "Stability and change in the Canadian central place system between 1971 and the early 1980s," in Nagle, F.N. (ed.), *Beitrage zur Stadtgeographie II: Stadtentwicklung in Ubersee*, Hamburg: Mitteilungen der geographischen Gesellschaft in Hamburg, 77: 3-31.

Preston, R.E. (1987) "A reassessment of the structure of classical central place theory," in Yadev, C.S. (ed.), *Models in Urban Geography: Theoretical (Part A)*, New Delhi: Concept Publishing Co.

Richardson, H.W. (1985) "Input-output and economic base multipliers: looking backward and forward," *Journal of Regional Science*, 25: 607-61.

Scorrar, D.A. and Williams, M.H. (1976) "Manufacturing activity and the urban hierarchy," in Ray, D.M. *et al.*, (eds.), *Canadian Urban Trends*, Vol. 1, Toronto: Copp Clark, pp. 105-40.

Semple, R.K. and Green, M.B. (1983) "Interurban corporate headquarters relocation in Canada," *Cahiers de Geographie du Quebec* 27: 389-406.

Simmons, J.W. and Bourne, L.S. (1981) "Urban and regional systems-qua systems," *Progress in Human Geography*, 5: 420-31.

Smith, J. (1965) "The growth of central places as a function of regional economy and population," *Swedish Journal of Economics*, 67: 279-307.

Tiebout, C.M. (1962) *The Community Economic Base Study*, New York: Committee for Economic Development.

Thompson, J.S. (1982) "An empirical note on the compatibility of central place models and economic base theory," *Journal of Regional Science*, 22: 97-103.

White, R.W. (1974) "Sketches of a dynamic central place theory," *Economic Geography*, 50: 219-27.

Woldenburg, M.J. (1968) "Energy flow and spatial order: mixed hexagonal hierarchies of central places," *Geographical Review*, 58: 552-74.

CHAPTER 14

A SEPARABLE PROGRAMMING APPROACH TO UPDATING I/O TABLES USING THE RAS CRITERION: WASHINGTON STATE 1967 TO 1972*

Christian M. Dufournaud
Department of Geography
University of Waterloo
Waterloo, Ontario

Updating input-output (I/O) tables is not a new endeavour for academics. There exists a rich literature of growing sophistication on the twin problems of deriving a regional I/O table from a survey based national table, and of deriving an updated regional table, given an initial survey based table for a region. The accumulated evidence suggests that nonsurvey based I/O tables are difficult to predict with accuracy, although in this regard an operational definition of accuracy needs to be provided. This paper intends to contribute to the growing body of literature by modelling the RAS criterion in a separable programming framework, permitting by extension, the introduction of

----

* I would like to thank Professor William Beyers of the Department of Geography of the University of Washington in Seattle for providing the Washington State I/O tables. Dr. Don Jones of Oak Ridge National Laboratories and Professor G. Hewings of the University of Illinois made many valuable suggestions which I intend to pursue in subsequent papers. I was unable to incorporate Professor Hewing's suggestions given the time constraints. Also I would like to thank the Social Sciences an Humanities Research Council of Canada for providing me a Leave Fellowship which allowed me to find the time to prepare the above article. I am personally honoured to submit this paper to the retirement Festschrift in honour of Professor Peter H. Nash Sr. of the Department of Geography of the University of Waterloo.

a greater set of constraints than the traditional RAS approach permits.

The paper proceeds by reviewing the contributions and critically assessing their findings. The equations used in the separable programming, I/O table updating procedure, are elaborated in the next section. The final section presents the results of applying separable programming to the problem of updating I/O tables and compares the results to those obtained by other authors on the same data base.

PREVIOUS WORK

The history of updating I/O tables begins with a description of the RAS method and the paper by Matuszewski *et al.* (1964). A lucid description of RAS is given in Morrison and Smith (1974) that, " A is a matrix of input-output coefficients, and u,v, and x are vectors of regional intermediate output, intermediate input, and gross output, respectively" (Morrison and Smith 1974, 9). The A matrix is assumed to be the input-output coefficient matrix which prevails at the national level. The RAS approach to obtaining a regional table is then structured by Morrison and Smith (1974) as follows:[1]

$$(1) \quad u_1 = A_1 x \,,$$

$$(2) \quad A_2 = \underline{u}\underline{u}_1^{-1} A_1 \,,$$

$$(3) \quad v_1 = i\underline{x}A_2' \,,$$

$$(4) \quad A_3 = A_2 \underline{v}\underline{v}^{-1} \,.$$

Underlined terms indicate diagonal matrices and i represents a vector of 1's. The process of solving for the regional coefficient table starts by inserting in equation (1) the national table and systematically solving the equations in succession. The process is repeated until u_1 and v_1 have stabilised at the regional values.

---

[1] A substantially similar notation to that used by Morrison and Smith (1974) is used in equations (1) to (4).

The Matuszewski *et al.* (1964) problem is a straightforward one. Given an I/O table at some period *t(previous)*, and given ". . . knowledge of the true values of final demand, of total output, and of primary input of each of the industries in the year for which the adjustment is being made, . . ." (Matuszewski *et al.*, 1964, 205), they specify a linear programming approach to updating the I/O table. The data specified above allow them to compute the updated coefficient table from row and column totals of the transaction table in the year *t(update)*, for which the adjustment is desired.

The Matuszewski *et al.* (1964) problem formulation is interesting for two reasons. First, they update the coefficients rather than the flows, something which few other authors do, and second, they introduce a set of constraints which results in a large number of nonzero variables in their basis. Their formulation is given as follows:[2]

$$(5) \quad \sum_i \sum_j |\overset{*}{a}_{ij}/a_{ij}-1| = \min$$

subject to:

$$(6) \quad \sum_i \overset{*}{a}_{ij} X_j^{56} = X_{.j}^{56} \quad \textit{for all } j,$$

$$(7) \quad \sum_j \overset{*}{a}_{ij} X_j^{56} = X_{i.}^{56} \quad \textit{for all } i,$$

and,

$$(8) \quad 1/2 \leq \overset{*}{a}_{ij}/a_{ij} \leq 2 \quad \textit{for all } i \textit{ and } j.$$

The coefficient with the star is the unknown one and the X variable on the left side of constraints (6) and (7) is the total output in the year *t(update)*. The right hand sides of constraints (6) and (7) are respectively total intermediate input into industry j and total intermediate sales of industry i in the year *t(update)* (Matuszewski *et al*, 1964, 205).

----

[2] The notation of Matuszewski *et al.* (1964) is reproduced in equations (5) to (8).

Clearly, the authors are minimising the difference between the transaction coefficients in one year and the transaction coefficients in another year subject to the updated row and column totals on transactions and subject to the fact that the coefficients cannot vary by more than 1/2 their value downwards, and by more than twice their value upwards. The latter upper and lower bounds on the change in the coefficients introduce more nonzero variables into the basis than would be required if only constraints (6) and (7) were used, a fact which brings a semblance of reality to the updated table since a typical I/O table usually contains more nonzero coefficients than twice the number of sectors minus one.

One problem not discussed by these authors is the nature of the data required to formulate their updating procedure successfully. The degree of confidence with which total final consumption, total intermediate inputs and total output are known is often small. Supposing that the data required by them was known with a low degree of confidence, their formulation allows the restating of constraints to reflect the lessened degree of certainty which one may have about the preliminary data requirements. The process of reformulating constraints to reflect a higher degree of uncertainty involves the relaxation of constraints. This process, however, in turn places the onus on the objective function to accurately capture the process of changing I/O coefficients, since the relaxation of constraints makes more solutions to the linear program at hand than previously was the case, a reality.

Simply minimising the change in coefficients required to remain consistent with known row and column totals as Matuszewski *et al.*, (1964) have done is clearly not appropriate in situations where the solution space is widened because of uncertainty about the values which the updated coefficients are to assume. It is possible to make assumptions about the direction in which coefficients will change when one passes from national to regional models or when one passes from a regional model in one time period to a regional model in another time period. For example in the latter case it might be possible to argue that a regional economy in the extreme case, had not experienced any change in sectoral production functions over time. If, however, change did occur, then that change would be in the direction of growing efficiency, implying a reduction in the input requirements per unit of final output for any given sector. Constraints forcing all updated coefficients to be no greater than the initial ones, could be added easily to the Matuszewski *et al.*, (1964) linear program by simply modifying constraint set (8).

It is an unusual logic, however, which based on the argument that there existed a degree of uncertainty about the change of row and/or column totals at the update period led to the introduction of a more binding set of constraints than those shown in equation set (8), based on assumptions about the direction of the likely change of the coefficients. A more sound approach relies on the structure of the objective function to yield updated coefficients.

Several candidates for an objective function which is more plausible than a linear one are available in the literature. The first is derivative from the RAS procedure and when reformulated by Theil (1967) is:

$$(9) \quad \min \sum_i \sum_j a^*_{ij} \ln(a^*_{ij}/a_{ij}) \; .$$

It is presented as shown above in several places in the literature starting with Theil and continuing with Hewings (1971), Morrison and Thumann (1980) and Harrigan and Buchanan (1984).

Criterion (9) can be better understood when it is portrayed as the negative of the entropy criterion. Indeed RAS minimises the negative of the entropy criterion or implicitly, maximises entropy. This means that RAS minimises the information gain required to be consistent with known row and column totals. Equation (9) is a convex function which achieves a minimum value for any unknown at the following point:

$$(10) \quad a^*_{ij} = a_{ij}/e \text{ for any } i \text{ and } j$$

where e is the base of the natural logarithm system. Equation (10) demonstrates that when criterion (9) is minimised, the predicted coefficients will be lower than the initial coefficients, consistent with the row and column constraints.

An alternative to criterion (9) is given in the literature by Morrison Thumann (1980) and Harrigan and Buchanan (1984) as:

$$(11) \quad \min \sum_i \sum_j (a_{ij} - a^*_{ij})^2 .$$

Criterion (11), a quadratic criterion, is also a convex function and achieves a minimum for any unknown at the following point

$$(12) \quad a^*_{ij} = a_{ij} \text{ for any } i \text{ and } j.$$

Unlike criterion (9), therefore, equation (12) demonstrates that when minimised, the quadratic criterion (11), yields coefficients which are equal to the initial ones, consistent with known row and column totals.

Which values are inserted into the quadratic criterion by both Morrison and Thumann (1980) and Harrigan and Buchanan (1980) is difficult to determine. Careful reading of their respective papers suggest that both sets of authors insert the actual intersectoral flows of goods and services rather than the coefficients of goods and services per unit of final output and that both also predict the actual intersectoral flows of goods and services rather than the unit sectoral flows. A rationale for predicting intersectoral flows rather than coefficients may be that total sectoral output is often unknown. Both sets of authors also introduce weighting schemes as follows.

$$(13) \quad \min \sum_i \sum_j (x_{ij} - \overset{*}{x}_{ij})^2 / \pi .$$

where π can take on a value equal to x_{ij} or x_{ij}^2 or $\sqrt{x_{ij}}$ for any i and j. Note that in equation (13) the unit coefficients have been replaced with actual intersectoral flows, x_{ij} .

Morrison and Thumann (1980), utilising the Washington State I/O tables for 1963 and 1967 at the 27 sector level of aggregation, minimise criterion (13) by Lagrangian multiplier methods subject to known row and column totals. They predict the 1967 table on the basis of the 1963 table, and the 1972 table on the basis of the 1963 table and on the basis of the 1967 table. The purpose of the exercise is to produce an updated table which is more accurate than the one which traditional RAS methodology yields in the same cases. To compute the success of their procedure, Morrison and Thumann (1980) utilise the average absolute error and the standard deviation of the absolute error as measures of deviation from the actual distribution of sectoral flows. Although it is not explicitly shown in their paper, the absolute error is calculated as:

$$(14) \quad \textit{absolute error} = |a_{ij} - \overset{*}{a}_{ij}| .$$

They also provide histograms of the distribution of the relative errors, computed as:

$$(15) \quad \textit{relative error} = (a_{ij} - \overset{*}{a}_{ij}) / a_{ij} .$$

Why Morrison and Thumann (1980) chose to compute the average absolute error and the standard deviation of the absolute error rather than the average and standard deviation of the error is not clear. It is obvious that both error distributions will have the same standard deviation but the average of the absolute errors will be higher than the average of the errors unless all values are underpredicted.

The Lagrangian approach which Morrison and Thumann (1980) present, lends itself to the introduction of more information about the structure of intersectoral flows than the row and column totals of the transaction table alone offers. Indeed, experiments which they conduct with the Lagrangian approach and systematically introducing more information about the transaction table, yield predictions of the actual values which are increasingly close to results obtained by the RAS approach. A drawback of the Lagrangian approach is that does not guarantee that only intersectoral flows greater than or equal to zero will be produced. A less serious additional drawback is that this approach requires the development of a specialised matrix generation routine.

Not surprisingly, Morrison and Thumann's (1980) results indicate that the greater the information entered as constraints into the Lagrangian procedure, the greater the accuracy of the predictions. A problem with their approach is that they provide no real evidence of the merits of the weighted quadratic criterion they are minimising relative to some other nonlinear criterion. A footnote on page 280 of their article cites Theil (1967) and shows that the weighted quadratic term is ". . . a first order approximation minimand. . .," (Morrison and Thumann, 1980, 280) of the RAS criterion. On page 281 they state: "Unfortunately, there is no generally agreed objective function at the present time. However, given the problem as defined, a feasible objective function seems to be the minimisation of the weighted squared sums of deviation. . ." Unfortunately their evidence overwhelmingly suggests that RAS produces better results[3] than the Lagrangian approach when only row and column constraints are used to solve both problems. It should also be observed that the first derivatives of the quadratic criterion yield linear equations, in turn providing that the solution of Morrison and Thumann's "constrained matrix problem" is amenable to solution via known methods for solving systems of linear equations.

----

[3] See their Table 1 and Table 4 on pages 284 and 286 respectively.

Morrison and Thumann do not study why their predictions using both RAS and the Lagrangian approach are not very accurate. One such factor is inflation. Inflation can impact the intersectoral flows in many ways. At the very least, even if a regional economy had not undergone any changes in its production functions and had not increased or decreased its sectoral output over time, but had experienced inflation, the intersectoral flows will be nominally higher amounts. In this regard it should be observed that the Washington State I/O tables are produced in current dollars.

Assuming that the Lagrangian approach of Morrison and Thumann was used to predict the coefficient values rather than the actual intersectoral flows, inflation would still pose a problem. Inflation impacts different sectors of the economy differently and for those sectors having been impacted less from inflation than other sectors, their coefficient flows to other sectors would *ceteris paribus*, actually decline as opposed to sectors having experienced high inflation for which their coefficient flows would rise.

Harrigan and Buchanan (1984) also use a quadratic criterion to predict updated I/O tables and regional I/O tables. They demonstrate that the weighted quadratic criterion is equivalent to the second term of a Taylor series expansion of the RAS criterion. They use quadratic programming and recommend this approach for three reasons. First, of these is that, ". . . it allows for the possibility that exogenous data are not known with certainty" (Harrigan and Buchanan, 1984, 342). The quadratic programming approach allows the row and column constraints to be relaxed from strict equality as would be required if the traditional approach to obtaining predictions under RAS was used. Harrigan and Buchanan show that it is possible to treat the right hand sides of the row and column constraints as initial values around which a weighted quadratic deviation function can be built and minimised along with quadratic functions built around each prospective intersectoral flow term.

A second reason Harrigan and Buchanan (1984, 343) advanced in support of a quadratic criterion" . . . is that it permits a wider variety and volume of information to be brought to bear on the updating process than is possible with the generalised RAS method." This is an obvious advantage of quadratic programming versus the traditional RAS method.

Lastly Harrigan and Buchanan (1984, 343) rest the merits of their weighted quadratic formulation on the fact that ". . . it nests many alternative formulations as special cases." The remainder of their paper deals with the application of the method of Hildreth to solving the quadratic programming problem and the application of this methodology to two cases. The first problem involves predicting the 1967 Washington State table based on initial values derived from the 1963 table, and the second problem involves using the U.K. table to predict the Scottish table.

The updating procedure of Harrigan and Buchanan is used on the actual intersectoral flows, similarly to Morrison and Thumann's. They offer little consideration of factors such as inflation, which could contribute to the imprecision with which predictions are produced. Their results also lead to the conclusion that traditional RAS methodology is superior than weighted quadratic programming.[4] Also of the three quadratic solutions Harrigan and Buchanan propose, distinguished by the particular weight used, the reader is not told which one was used for the comparisons appearing in their Table 2 on page 355. It must be presumed that the solution yielding the best values i.e. values closest to the actuals, is used, but it would be useful to note which weighting scheme was associated with the closest solution to the actual data. If, for example, it were the one whose weights bring the quadratic criterion closest in line with the second term of the Taylor series expansion of the RAS criterion as defined by Theil (1967), this would constitute preliminary evidence that the closer the criterion is to RAS, the better job it does of predicting I/O intersectoral flows.

LTERNATIVE OPERATIONAL FORMULATIONS OF THE PDATING PROCESS

The previous. section reviewed four different updating procedures. I argued that none of the authors discussed factors such as inflation, which could explain why the predictions provided in the updates do not conform more closely to the actual values. Morrison nd Thumann's (1980), results suggest that RAS is superior to the agrangian approach, while Harrigan and Buchanan's (1984) present nsufficient evidence to judge which weighting scheme used in the quadratic prediction comes closest to predicting the actual values. There is, however, enough evidence to suggest that RAS still does a better job of predicting actual values than any of the quadratic criteria.

---

[4] See Table 2 Harrigan and Buchanan (1984) page 355.

The analysis developed in this section is applied to the problem of updating the 1967 Washington State I/O table to 1972 and comparing the results to the actual 1972 intersectoral flows at the 27-sector level of aggregation. The approach developed solves the RAS criterion as reformulated by Theil (1967), as a separable programming problem subject to the row and column constraints and other constraints reflecting any additional information available about the problem at hand. This can be accomplished as follows:

$$(16) \quad \min \sum_i \sum_j \left[X_{ij}/e\{\ln(1/e)\}(\pi_{1,ij}) + n(X_{ij}\{\ln(n)\})(\pi_{2,ij})\right],$$

subject to:

$$(17) \quad \sum_j \left[X_{ij}/e(\pi_{1,ij}) + n\{X_{ij}(\pi_{2,ij})\}\right] = X_{i.} \quad \textit{for all } i,$$

$$(18) \quad \sum_i \left[X_{ij}/e(\pi_{1,ij}) + n\{X_{ij}(\pi_{2,ij})\}\right] = X_{.j} \quad \textit{for all } j,$$

and,

$$(19) \quad \pi_{1,ij}+\pi_{2,ij} = 1 \textit{ for all } i \textit{ and } j.$$

Equations (16) to (19) require explanation. The starting point should be equations (9) and (10), the latter showing where the RAS criterion achieves a minimum. Equation (16) states that for every previously known intersectoral flow greater than zero, the value to be predicted is allowed to vary between the value that makes criterion (9) a minimum and n times the previously known intersectoral flow. The two probabilities $\pi_{1,ij}$ and $\pi_{2,ij}$ are the unknowns which interpolate any value between the above described extremes. These probabilities must sum to one according to equation (19). Equations (17) and (18) state that the sum of the intersectoral flows, expressed in a manner to be consistent with the separable programming framework, must equal the known future row and column totals.

The problem as formulated in equations (16) to (17) leads to rounding errors which cause infeasible solutions. The problem can be modified to circumvent rounding error by first defining the left hand sides of equations (16), (17), and (18) as being equal to R_0, R_i and R_j and then redefining the separable program as:

$$(20) \quad \min R_0 + \sum_i (M^+ d_i + M^- d_i) + \sum_j (M^+ d_j + M^- d_j),$$

subject to:

$$(21) \quad R_i + {}^+d_i + {}^-d_i = X_{i.} \; \textit{for all } i,$$

and,

$$(22) \quad R_j + {}^+d_j + {}^-d_j = X_{.j} \; \textit{for all } j.$$

Equation set (19) does not need to be modified. M is made sufficiently large for it to be minimised before any other terms in the objective function.

RESULTS OF THE SEPARABLE PROGRAMMING PROCEDURE

In updating the 1967 Washington State table to 1972, the 436 with nonzero flows of the 1967 table are used to create a problem with 872 probability variables and 54 times 2 = 108 variables of type d, for a total of 980 variables. There are 27 constraints associated with equation (21), 27 constraints associated with equation (22) and 436 constraints associated with equation (19) for a total of 490 constraints. In addition to the problem as formulated above, the possibility of inflation is introduced by inflating the previously known intersectoral flows. Inflation is assumed to take on the values 0, .1 and .2 and is assumed to affect each intersectoral flow equally. Finally, n is allowed to assume the values 2, 3 and 5. Thus, a total of nine different problems are solved with the above parameters. Three additional problems, one for each assumed inflation value, where the predicted values are allowed to vary only between the initial intersectoral flows and 2 times the initial intersectoral flows, are also solved. The results of these twelve solutions are presented in Table 1.

Contrasting the results of Table 1 with the results of Morrison and Thumann (1980) for the identical updating problem, reveals that for all updates produced by separable programming, for which the value of n is set at 2 or 3, the standard deviation of the errors is smaller than the standard deviation resulting from applying the

Table 1

Frequency Distributions of Errors Under Different Assumptions about the Minimum, Maximum and Inflation Values

	Absolute Values of Relative Error Size in %-ages	M/0/2	M/.1/2	M/.2/2	M/0/3	M/.1/3	M/.2/3	M/0/5	M/.1/5	M/.2/5	I/0/2	I/.1/2	I/.2/2
1	0-0.9	7	2	0	4	1	0	1	1	1	7	1	0
2	1-4.9	7	10	4	5	8	5	1	2	5	6	11	7
3	5-9.9	16	2	10	8	6	5	4	4	3	12	3	6
4	10-19.9	16	34	18	21	23	23	8	13	11	15	30	17
5	20-49.9	88	87	87	58	61	54	50	45	36	87	83	86
6	50-99.9	189	188	198	207	194	203	220	218	230	221	220	240
7	100-199.9	112	112	112	110	115	118	105	111	93	99	98	94
8	>=200	44	44	50	66	71	71	90	85	100	32	33	29
	SUM	479	479	479	479	479	479	479	479	479	479	479	479
	Mean Error	6.21	6.50	6.79	7.28	7.54	7.77	8.08	8.18	8.25	4.95	4.91	4.89
	St. Dev.	20.56	22.71	24.19	25.77	25.79	26.62	31.48	30.96	33.47	17.00	16.81	17.43

M or I = minimum value or initial value / inflation value / value of 'n' as described in text traditional RAS procedure.[5] Increasing the value of n to 5 causes the standard deviation of errors derived from separable programming, to exceed the standard deviation of errors produced by RAS.

Changing the inflation factor produces no noticeable reduction in the standard deviation of errors. This should not lead to the conclusion that inflation is not an underlying factor causing the updating procedure to predict results which are different from the actual values. In the current analysis, a single inflation value was used to inflate all initial values, clearly an oversimplification of the real process. Ordinarily it can be expected that inflations affects different sectors differently. Moreover, the actual values used, 0, .1 and .2, might not have captured the nominal change which inflation caused in the State of Washington between 1967 and 1972. It might be that 0.5 is on average, the value which should have been used and that therefore the analysis reported here missed the true value.

The frequency distributions are disconcerting. Around three-quarters of the true values are at least fifty per cent larger or smaller than those predicted. A consoling observation is that when the predicted values are allowed to vary only between the initial values and two times the initial values, the number of true values that are at least 100 per cent different from the predicted ones declines significantly.

A number of subsequent analyses are possible. Firstly, rather than predict the actual intersectoral flows, the coefficient flows should be predicted. These latter terms are derived from the former by dividing by gross sectoral output. This division eliminates one source of variation between coefficients in two different years, i.e. the variation due to changes in gross output. This leaves changes in the production function and changes due to inflation as possible sources of variation in the coefficients between two years. As Harrigan and Buchanan (1984) note, quadratic criteria are amenable to solution by separable programming. Thus, follow up analyses should compare the RAS criterion to the weighted quadratic when both are solved by separable programming, and in situations where additional information is known about the I/O table, to be predicted. Inflation should be treated in a more realistic manner than the preliminary treatment it received in this paper. Sectoral inflation rates should be obtained and these applied to their respective sectors.

---

[5] See Morrison and Thumann's Table 1, page 284.

Equation (10) shows that the RAS criterion achieves a minimum value at a smaller predicted coefficient than the quadratic criterion whose minimum is shown in equation (12). This suggests that an economy having made gains in efficiency, i.e. having lowered its input coefficients for some sectors over time, would be predicted more accurately by the RAS criterion than a quadratic one.

In analysis presented here, only the nonzero flows in the initial year were assumed to have nonzero flows in the predicted year. This is clearly nothing more than an assumption useful to reduce the number of variables to be included in any analysis. Future analysis could model the initially known nonzero flows using the RAS criterion simultaneously to modelling the initially known zero flows using quadratic terms. In a separable programming framework, the initially known zero terms could be allowed to vary between a minimum of zero and a maximum value based on the average size of small intersectoral coefficients in the initially known table. Introducing the possibility of allowing some coefficients to change from zero to a number greater than zero might permit the d variables to be removed, since these same coefficients would act as slack variables. The equality constraints on equations (17) and (18) probably would need to be changed to greater-than constraints in that event.

CONCLUSIONS

A number of techniques used to update I/O tables were presented and some updates of the 1967 Washington State table to the 1972 table were presented. This preliminary analysis, while disappointing, has been fruitful enough to make follow up analyses as suggested in the previous section worthwhile.

REFERENCES

Harrigan, F. and Buchanan, I. (1984) "A quadratic programming approach to input-output estimation and simulation," *Journal of Regional Science*, 24: 339-58.

Hewings, G.J.D. (1974) "Regional input-output models in the U.K.: some problems and prospects for the use of nonsurvey techniques," *Regional Studies*, 5: 11-22.

Matuszewski, T. Pitts, P.R. and Sawyer, J.A. (1964) "Linear programming estimates of changes in input-output coefficients," *Canadian Journal of Economics and Political Science*, 30: 203-10.

Morrison, W. and Smith, P. (1974) "Nonsurvey input-output techniques at the small area level: an evaluation," *Journal of Regional Science*, 14: 1-14.

Morrison, W. and Thumann, R.G. (1980) "A Lagrangian multiplier approach to the solution of a constrained matrix problem," *Journal of Regional Science*, 20: 279-92.

Theil, H. (1967) *Economics and Information Theory*, Amsterdam: North Holland.

CHAPTER 15

SOVIET URBAN DEVELOPMENT: FIRST PRINCIPLES AND PRESENT REALITIES

James H. Bater
Dean, Faculty of Environmental Studies
University of Waterloo
Waterloo, Ontario

The Soviet Union is a land of large cities as Chauncy Harris (1970, p. 1) noted in his classic study, *Cities of the Soviet Union*, published nearly two decades ago. Little that has happened since would alter that judgment. The number of people living in small cities and urban-type settlements of less than 50,000 inhabitants has certainly increased since 1970, but so too has the number living in places of more than one million. Moreover, the share of the total urban population represented by the former group has declined, while the latter has grown. And both trends are in stark contrast to long-standing Soviet policies intended to guide urban development. This paper will review the evolution of some of the more important urban policies and assess their impact on the distribution of urban places. Finally a few tentative ideas as to urban development trends during the Gorbachev era will be offered.

The definition of what is urban in the Soviet Union is by North American or European standards rather generous, if one anticipates by such a definition a place that appears urban. Cities in the Soviet Union are currently defined as those places with more than 10,000--12,000 inhabitants, the actual number varying according to republic. It is fair to say that in the case of an agricultural or resource development settlement recently promoted to city rank there is often little evidence of urban attributes, namely, an identifiable core, a reasonable range of consumer and cultural services, or even adequate basic municipal services such as water, sewer and paved roads. This is even more apt to be the case for those places defined as urban-type settlements. With between 2,000 and 3,000 people and 60-85 per cent of the labour force engaged in nonagricultural occupations, such places often are urban in only the most rudimentary sense (Shabad, 1984, p. 96). To be sure, the urban scene has been radically transformed during the Soviet era (Bater, 1980). After all, at the time of the revolution barely one-sixth of the total population was classified as urban, and during the chaos of the civil war years immediately following the assumption of political power by the Bolsheviks urban-rural migration reduced this even small percentage. Since the late 1920s, however, rapid urban-industrialisation has produced fundamental changes. More than 1,200 new cities have been created, and at present nearly two-thirds of the 280 million total population is urban. But it should be remembered that it was only in 1961 that the urban population exceeded the rural (Figure 1). In a variety of ways the link with the countryside remains strong for many citizens of the present-day Soviet city.

FIRST PRINCIPLES

The Soviet socialist city heralded new opportunities for charting the course of urbanisation. With all resources nationalised, land uses everywhere were to be planned. And with the prospect of controlling the urban growth process, there was the possibility of removing the inherent contradiction between city and countryside which Marxists had long argued was a by-product of the private ownership and unplanned development of resources under capitalism. While not much was actually built during the 1920s owing to the deteriorated national economy and general chaos of the times, debate over what form the new Soviet socialist city should take, and what the geography of settlement ought to look like in the future, was intense. Amongst the many schools of thought two stand out as representative. They are most often labelled the urbanist and de-urbanist approaches.

The urbanists argued that a system of self-contained urban centres would serve well the needs of the socialist future. In such a system of cities state control over the size of the population and multistoreyed collective living facilities which deemphasised distinctions between the various strata of Soviet society would be the key to creating a communal ethos. Strict land use zoning, predominantly pedestrian movement, a host of state services such as day-care, communal food services and so forth would facilitate the realisation of some of the other basic goals of socialism, such as the betterment of the urban environment and the dissolution of the nuclear families that resided within it. On the matter of eradicating

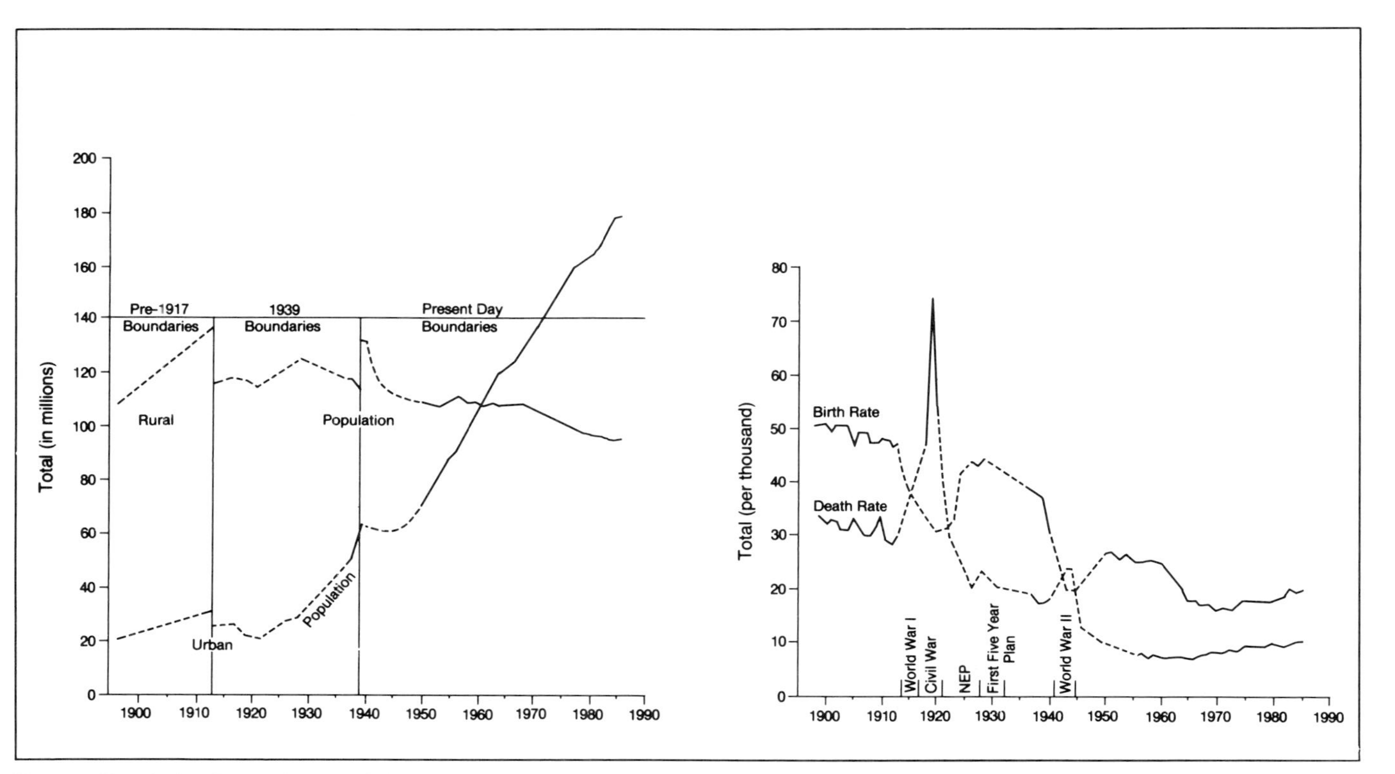

Figure 1: Population Trends in the U.S.S.R. 1900-1986

the differences between city and countryside the urbanists had little to say other than that Soviet socialist society should eventually be predominantly urban. On this point the de-urbanist school was far from silent.

The de-urbanists contended that the urbanists instead of charting a new and socialist urban future were simply proposing to replace large cities with smaller ones. The de-urbanists, on the other hand, wanted to create an entirely new settlement form, one more in harmony with what they perceived as basic Marxist ideology. They argued for an essentially townless, socialist society. In it the population would be distributed over all of the habitable areas of the state in a ribbon-like pattern, thereby ending the age-old tension between urban and rural. Within such settlements individual dwellings would be located in natural surroundings within easy reach of communal centres for dining, recreation, health care, education and so on. While some measure of privacy would be provided in terms of living arrangements, the lifestyle of the inhabitants would be essentially communal. All centres of employment and consumer services would be situated in these same ribbon-like developments, and as in the urbanist scheme, in such a manner to minimise travel time to them. But spatial mobility was predicated on universal use of the automobile, in marked contrast to the urbanists' preoccupation with pedestrian movements (Bater, 1980, pp. 21-31).

Given the radical nature of the planning proposals put forward by the urbanist, and especially the de-urbanist schools, it is scarcely surprising that few schemes ever got past the drawing board (Starr, 1978). As was so often the case during the early Soviet period, visionary, utopian and not uncommonly, totally impractical experimentation had little impact on ordinary Soviet citizens. The vast majority were still villagers whose daily lives were often guided by practices more medieval than modern. When Stalin initiated a program of forced industrialisation with the First Five Year Plan in 1928, the long-term development priorities of the state were very clearly laid out. A fundamental restructuring of the existing urban system was simply not included.

The debate over the future form of the Soviet socialist city ended in 1931 when it was decreed that all Soviet cities must be socialist by virtue of their being part of the Union of Soviet Socialist Republics. This pragmatic policy statement came with the full authority of the Communist Party. Thus it served to stifle further public debate, and especially that centred on the complete eradication of the existing urban system which was the legacy of the imperial, quasi-capitalist era. However, this did not mean that the ideas put forward during the 1920s were without consequence. In fact many of the basic concepts underlying the urbanist and de-urbanist positions were incorporated into the Plan for the Reconstruction of Moscow adopted in 1935, albeit in suitably modified form. Among the more important were: a limited size of population; an ideological role for the city centre; state control and allocation of housing; spatial equality in the distribution of consumer and cultural services; a limited journey to work. The notion of an optimal size city was endorsed. A population of between 50,000 and 60,000 was deemed to be ideal. For Moscow, which already had more than three million citizens, the ultimate limit was set at five. Clearly, what the Moscow Plan of 1935 heralded was scarcely the utopia that planners and architects of the 1920s envisioned.[1] Nonetheless, the principles underlying the plan did constitute a sharp break with a past practice so often based upon simple expediency.

Over the years the general principles for urban development have been translated into myriad norms to guide planners. Population growth rates were established, housing allocations were standardised, maximum times for the journey to work according to city size were set, the level of consumer service provisionment was calculated on a per capita basis, and so on. While there was relatively little by way of explicit guidelines for the development of the urban system, that there must be a systems perspective was recognised. A 1933 regulation stipulating that new towns could not be planned in isolation is one indication of this (Andrusz, 1984, p. 345, ref. 111). While planning principles and the determination of operational guidelines are one matter, implementation of them is clearly quite another. At this point it would be useful to describe briefly the principal mechanisms available to the state for the purpose of manipulating the movement of people.

MECHANISMS FOR MANIPULATING POPULATION MOVEMENT

The Soviet version of the internal passport introduced in 1932 was scarcely a novel concept. A similar device was employed by the Tsarist government for essentially the same purpose, namely, to control the movement of peasants from village to city. For those aged 16 or older leaving the countryside for the city without a passport during the Stalin era was not impossible, but it certainly was difficult.

---

[1] For related comments see Bater, 1984.

For the millions of adult *kolkhozniki*, that is, collective farm peasants, who comprised the bulk of the rural population, it remained so for decades. It was not until the mid-1970s that they were legally entitled to an internal passport upon reaching age 16. It is only since 1981 that all Soviet citizens aged 16 and over have been required to have an internal passport. This document facilitates movement, both in terms of permanent relocation and holidaying. In the former case it is a necessary document in seeking employment, and in the latter it is usually requested when purchasing airline and rail tickets. Aside from establishing identity, the passport also provides authorities with information on place of residence and nationality. Clearly, monitoring, if not actually controlling, internal population movement is more easily managed by virtue of the passport system. But residence in a city is not assured simply because one possesses a passport.

The second bureaucratic device for controlling urban growth is the *propiska*, or residence permit. Each urban resident must have a *propiska*, which is entered in the passport and which gives legal recognition of the right to live in a specific city. The *propiska* denotes whether residence is permanent or temporary. Temporary residence permits of between one and three years have been quite common owing to the severe labour shortage in some cities, notably the larger ones. Currently, people holding temporary *propiski* are reckoned to comprise about 15 per cent of the labour force in the large Soviet city (cited in Loeber, 1984, p. 304). Often referred to as *limitchiki* because of their temporary urban status, they represent a kind of legal lumpenproletariat. In theory, one cannot live in a city without a *propiska*, but in fact over the years most Soviet cities have attracted sizeable numbers of "illegal" unregistered residents. Registered migrants are normally permitted a three-month temporary *propiska* during which time they must secure a job if they arrived without one in hand. When credentials and employer's recommendation are in order, migrants may then obtain a permanent *propiska*, which allows them to be assigned accommodation. Given the general shortage of housing, prolonged waiting times are not unusual. For many this means that home continues to be a bed in a hotel, hostel, workers' barracks, a friend's apartment or even rented space in someone else's home. It requires little imagination to see how internal migration and urban growth might be "administratively" manipulated through discretionary allocation of an internal passport and *propiska*. However, a lot of imagination and enterprise have been employed in overcoming such institutional barriers to internal migration.[2]

As well, the state is theoretically able to manipulate the urban growth process through the discretionary allocation of investment funds, especially for industry. Directing investment to specific regions, or to specific size groups of cities, while at the same time restricting investment in particular places, has long been part of Soviet urban policy.[3] As early as 1931 decrees were issued which sought to restrict investment in Moscow and Leningrad to those activities which served the existing population. Combined with controls over in-migration the growth of the country's two largest cities was thereby to be regulated. By 1939 this policy of restricting investment had been extended to the industrial centres of Khar'kov, Kiev, Rostov-on-Don, Gor'kiy and Sverdlovsk. By 1956 a prohibition on construction of new industry and expanding existing plants applied in 47 cities. In 23 others industrial investment was to be limited.[4] In the early 1980s similar restrictions applied in about 200 cities, including the centres in the Moscow and Leningrad metropolitan areas. All told, industrial investment is either prohibited or limited in roughly two-thirds of all the cities with more than 100,000 inhabitants. Along with the policy of curtailing industrial investment is that which directs ministries to locate new facilities in the small, more remote urban centres, especially those with fewer than 50,000 population (Rowland, 1980). Such policies, and the mechanisms available for their implementation, provide the basis for the conscious manipulation of the Soviet population fully in accord with first principles. The relationship between principle and practice is, of course, not necessarily a very close one.

---

[2] Obtaining a *propiska* through an arranged marriage is a not uncommon ploy. For further discussion see Zaslavsky, and Luryi, 1979, pp. 137-53.

[3] An overview of methods used is provided by Khodzhayev, 1986, pp. 12-14.

[4] Khorev lists 48 cities while Rowland uses 47 in his analysis Khorev (1986, p. 27, Rowland, 1986, pp. 653-54).

FIRST PRINCIPLES AND PRESENT REALITIES

It has been implied that in a Soviet type command economy it is possible, in theory at least, to manipulate population movements to conform with state policies. This presumes concurrence on the part of the people being "manipulated." But it will come as no surprise that when state policies do not match individual or interest group preferences, state policies are not infrequently confounded. The Soviet experience in manpower planning is instructive in this regard and therefore some features of internal migration during the Soviet period will be examined briefly. With this overview in hand, the discussion will turn to a consideration of the management of urban growth by city-size group.

In simple terms the major population movements during the Soviet period have been from west to east and from north to south. Since the turn of the century the absolute population increase east of the Volga River has been about twice that west of it. To be sure, part of this growth reflects the comparatively high per capita crude rates of net natural increase which differentiate the Middle Asian and Caucasian peoples who predominate east and south of the Volga from the Slavs and Baltic peoples of European Russia. But a substantial component of this population increase is the result of internal migration, both voluntary and involuntary. Development of the huge resource potential of the eastern regions was a major objective of the Stalin-era industrialisation drive. There has continued to be a commitment to develop the eastern regions for most of the post-Stalin period (Dienes, 1984). However, reliance on incentives has assumed more importance in attracting labour to the eastern and northern frontiers than simply assigning people to work there for indefinite periods, as was commonplace with Stalin. Indeed, with the easing of the Stalin era restrictions on internal migration in the late 1950s, substantial numbers of people decided to leave the eastern regions. Between the 1959 and 1970 censuses net out-migration from the West and East Siberian economic regions alone was close to 900,000.[5] While the total population in these regions still grew as a result of natural increase, the scale of the outflow clearly posed some serious manpower problems. Thus, in the late 1960s regional wage incentives in the eastern zones were augmented. Combined with public appeals for Soviet youth to participate in the construction of massive projects

---

[5] For a detailed discussion of intercensal population change, see Lewis and Rowland, 1979.

such as the Baykal-Amur railroad (BAM), thousands of new workers have since been drawn into the eastern regions from other parts of the country. Thus, net out-migration which characterised the 1960s appears to have been stemmed for the time being (Sheehy, 1985, pp. 2-4). The problem confronting Soviet manpower planners now is that despite sizeable wage differentials the urban-industrial centres in the eastern regions are generally characterised by very high labour turnover rates. In many industrial operations it exceeds 50 per cent per annum. Thus, not only does labour cost more than say in European Russia, there is the added burden of continually training new workers whose productivity is naturally much lower than seasoned employees. While many of those who migrate to the eastern frontiers do not stay long, nonetheless one of the major achievements of the Soviet era is the quite fundamental re-working of the distribution of population.

The shift in the centre of gravity of population from north to south, while perhaps more subtle than from west to east, has been especially frustrating in terms of balancing supply and demand for labour. The Soviet equivalent of the sun-belt - Moldavia, the south Ukraine including the Crimea, the north Caucasus and Caucasian republics, and Middle Asia - not only is a region characterised by high rates of natural increase, over the years it has been the destination of a substantial number of migrants from the north. The majority of the migrants are Slavic and urban. For example, from the Russian Soviet Federative Socialist Republic (RSFSR) alone more than one million people left for a sunnier southern clime between 1970 and 1979. Slavic migrants traditionally have found employment in urban industry, and often dominated the managerial ranks. But in recent years there are signs of a change in direction of migration. As Table 1 indicates, between 1979 and 1984 the RSFSR recorded a net in-migration of nearly three-quarters of a million people. The data in Table 1 suggests that this is because of an out-migration from the southern republics. There are several possible reasons for this quite significant change in basic migration pattern. As Middle Asians themselves become more urbanised they are bringing pressure to bear on migrant worker and manager alike in terms of competition for jobs. And when indigenous people reach positions of power and authority there are often claims that they show favouritism toward their own kind, a scarcely surprising development but one which is resented by migrants. Ethnic tensions are conceivably another reason for the change in migration flow. The recent racial riots in Kazakhstan were widely reported in the Soviet press and certainly will

Table 1
Natural Increase and Balance of Migration in Union Republics Between January 17, 1979, and January 1, 1984

		of which	
	Increase in Population	Natural Increase	Balance of Migration
RSFSR*	4,566,000	3,819,000	+ 747,000
Ukraine	912,000	917,000	- 5,000
Belorussia	318,000	323,000	- 5,000
Moldavia	133,000	207,000	- 74,000
Lithuania	141,000	89,000	+ 52,000
Latvia	66,000	25,000	+ 41,000
Estonia	52,000	24,000	+ 28,000
Georgia	152,000	240,000	- 88,000
Azerbaijan	478,000	582,000	- 104,000
Armenia	236,000	274,000	- 38,000
Uzbekistan	2,107,000	2,218,000	- 111,000
Kirgizia	357,000	413,000	- 56,000
Tajikistan	564,000	607,000	- 43,000
Turkmenistan	359,000	385,000	- 26,000
Kazakhstan	964,000	1,224,000	- 260,000

Source: Ann Sheehy, "Population Trends in the Union Republics, 1979-1984", *Radio Liberty Research Bulletin,* Vol. 29, No. 22, (3331), May 29, 1985, 7.

*Russian Soviet Federative Socialist Republic

serve to dampen Slavic enthusiasm to move to Middle Asia.[6] They are symptomatic of some fundamental problems and no doubt will spur even more Slavic migrants to return home. Still, it should not be forgotten that given the harsh winter which exists over much of the northern region the attraction of life in a southerly locale remains

---

[6] See "The riots in Alma-Ata: what caused them?" *The Current Digest of the Soviet Press*, 39(1) February 4, 1987, pp. 4-6.

compelling.

Thus, the paths of migration in recent years raise some new problems. If the pattern depicted in Table 1 continues then the Slavic presence in the Middle Asian and Caucasian realm will diminish, something scarcely desired by the Party or the government. The in-migration registered by the RSFSR would appear to be a positive development since this is the key labour deficit republic. But if it transpires that the principal destinations are the major cities of European Russia, the east and north will continue to be labour deficient. A return to the days of limited choice over job and its location, reminiscent of the early Stalin era in general and the war measures era in particular, is most unlikely. As restrictions on internal migration have eased since the late 1950s people have come to expect to exercise personal choice as a matter of course. The not uncommon pattern of migration from labour scarce northern and eastern regions to labour surplus southern locales is one instance of this attitude. The persistence young people in European Russia show in leaving the labour short villages for the town is another.

To some degree then migration has been managed in accordance with state policies. But there still are some unrealised objectives: for example, stabilising rural populations in marginal, but still important, agricultural regions in European Russia; moving indigenous Middle Asians from the labour surplus countryside to the city; promoting urban growth in particular regions and specific city-size categories. At this point it would be useful to review the Soviet experience in this latter regard.

The industrialisation drive initiated by Stalin in the late 1920s set in motion a seemingly endless circular and cumulative process of urban growth. Notwithstanding the control mechanisms enumerated earlier, planners were unable to hold city growth in check. Ideas about the optimal size city soon began to change. Originally conceived to be in the 50,000--60,000 range in the 1920s, three decades later the optimum sizes most frequently cited ranged between 150,000 and 200,000. This represented a pragmatic accommodation with the actual course of urban growth, but was also indicative of changing ideas as to the nature of urban economies. By the early 1960s the optimum size had been bumped into the 200,000--300,000 range. Put simply, despite the steady inflation of what constituted the optimal size for a Soviet city, the actual tempo of growth far outstripped it. By the beginning of the 1960s the validity of the concept was being widely questioned. Events since have put paid to its practical value in planning theory and practice.[7]

From Table 2 it is evident that the distribution of the urban population by city-size group has altered substantially since 1926. At that time almost half of the Soviet urban population lived in cities of less than 50,000 inhabitants. By the mid-1980s barely 30 per cent did. During this period a substantial number of cities joined the largest city-size category of over 500,000 population. This city-size group more than doubled its share of the total population between 1926 and 1985. Even when the specific cities in this largest city-size category at a particular time are taken as the basis of measuring rates of growth, invariably they grew at least as fast as the cities of less than 50,000 (Rowland, 1986, pp. 653-54).
In short, policies to check the growth of the largest urban places and to promote the expansion of the smallest have not been terribly effective.

Perhaps the clearest indication of the nature of the urban growth process during the past six decades is given by the number of cities with more than one million inhabitants. In 1959 there were only three: Moscow, Leningrad and Kiev. By 1970 seven more had been added to the roster. In 1986 there were 22 million cities and fully one-fifth of the urban population lived in them.[8] There is, of course, nothing magic about one million. However, the trends underlying the differential growth rates suggested by the data in Table 2 are significant. Population concentration in large cities was originally perceived to be a negative phenomenon, an inherent feature of capitalist societies where urban growth was both spontaneous and uncontrolled. It is still the view of many Soviets that "controlled" urban growth should be one of the distinctive features of a socialist system. Yet as we have seen, despite a flood of decrees establishing limits to city size, the existence of a system of internal passports and *propiski*, and a real need for planners to predict city populations, urban growth and especially that of the larger centres, has proceeded at largely unplanned rates. The consequences clearly have been prejudicial to the planned development of the urban system. But why has this situation arisen? One of the principal reasons is the pattern of investment in industry.

----

[7] For a general discussion, see Jensen, 1984.
[8] *Narodnoye Khozyaystvo SSSR v 1985 g*, Moscow: Finansy i Statistiki, 1986, pp. 1, 18-23.

Table 2
Urban Growth by City Size Group
(in million)

City Size Group	Population											
	1926		**1939**		**1959**		**1970**		**1977**		**1985**	
	m	%	**m**	%	**m**	%	**m**	%	**m**	%	**m**	%
Less than 50,000	12.7	(48.3)	24.9	(41.2)	40.4	(40.4)	47.4	(34.8)	48.7	(30.5)	54.5	(30.3)
50,000 - 100,000	4.1	(15.6)	7.0	(11.6)	11.0	(11.0)	13.0	(9.6)	15.3	(9.6)	17.0	(9.4)
101,000 - 500,000	5.4	(20.5)	15.7	(26.0)	24.4	(24.4)	38.3	(28.2)	46.8	(29.3)	50.6	(28.1)
over 500,000	4.1	(15.6)	12.8	(21.2)	24.2	(24.2)	37.3	(27.4)	48.8	(30.6)	58.0	(32.2)
of which over 1,000,000	– N/A –	7.7	(12.7)	10.1	(10.1)	19.1	(14.0)	27.4	(17.2)	38.3	(21.2)	
Total	26.3	(100)	60.4	(100)	100	(100)	136.0	(100)	159.6	(100)	180.1	(100)

Source: *Narodnoye Khozyaystvo SSSR za 60 Let* (Moscow, 1977), 59-68;
Narodnoye Khozyaystvo SSSR v 1984 g (Moscow, 1985) 5, 20-25.

With ministries and their various departments and enterprises increasingly preoccupied with efficiency in the allocation of limited financial resources, linked closely to the issue of profitability introduced by the 1965 economic reform, and more recently endorsed by Gorbachev, the financial benefit accruing from the external economies of the large urban centre clearly put a premium on location there. For the enterprise seeking to expand *in situ* owing to the lower incremental cost than in a new location, a major stumbling block can be the city planning agency which may regard industrial expansion as undesirable. However, the city Soviet and its various regulatory and planning departments are frequently no match for politically and economically important ministries and enterprises. Thus, notwithstanding the numerous regulations restricting the growth of large cities, the existing deficiencies in housing, consumer and cultural services within them, and the obvious negative impacts an unplanned population increase would bring, expansion of the employment base in the large urban centre often goes ahead regardless. At present more than 40 per cent of investment in industrial production is concentrated in cities with more than 500,000 population (Listengurt, 1985, p. 70). Without more effective intervention this share is likely to continue to increase.

Even when civic authorities are seemingly successful in warding off unwanted increased employment in a particular production facility, there are other means to achieve the same end. If an enterprise is advised it cannot expand its employment, and hence the population of the city concerned, it is not uncommon for production to be increased anyway by simply attracting labour from *outside* the city. Of the 58 million people who commuted to cities in 1980, more than 17 million did so to work. This so-called "pendulum" migration has grown rapidly, and is having an impact on the city, the rural sector and on the nature of the settlement system itself (Demko and Fuchs, 1983, p. 550).

Many enterprises are easily able to draw upon a labour supply from outside the city because of the substantial improvement to the Soviet ground transportation system which has occurred over the past three decades. Modern commuter trains now degorge enormous numbers of people compared to the late 1950s. Improvements to the network of bus routes, and the frequency of service on them, have similarly enabled ever larger numbers of rural residents to travel to the city to work, shop or use services not otherwise available to them. Between 1960 and 1980 the total number of commuters increased more than twofold. At the latter date about 7.5 million people travelling each day to work in a city began the trip in a village. Roughly the same number commuted from one city to another.

In general the zone of influence varies according to the size of the city. In the case of Moscow, for example, it extended some 20 kilometres beyond the city border in 1920. It now encompasses an area at least 50 kilometres beyond the city borders, which in turn obviously encompass a vastly larger city than in 1920. Each day some 500,000 people pour into Moscow from surrounding urban and rural settlements, while perhaps as many as 100,000 depart the city to jobs in outlying enterprises. To be sure a substantial number of people commute long distances, spending as much as two hours each way because they are unable to obtain a *propiska* for the city in question. But to judge from recent surveys of job satisfaction, many commuters are choosing a "suburban" lifestyle. The attractions of village life are perhaps greater if one's income and lifestyle are not directly tied to the land (Demko and Fuchs, 1983, pp. 554-55). In this regard it is worth remembering that many urbanites have recently relocated from village to city. After all, it was only in 1961 that even half of the Soviet population was urban. Village life therefore remains important. Huge numbers of people regularly visit relatives in the countryside, or travel to the family home in the village, now serving the grander function of weekend *dacha*, or cottage. Whether for recreation, food supply or familial obligation, the connection between town and country remains strong for millions of Soviet urban citizens. The impact of such commutation on countryside customs and values is important. Thus, while controls over population movement limit entry to the major cities, surrounding communities are often experiencing exceedingly rapid growth.

Even in the case of Moscow where planning authority extends well beyond the "official" city border, the growth rate of settlements subordinate to the Moscow City Soviet is, for some observers, alarmingly high. Between 1959 and 1981 the population of Moscow itself increased from 5.1 to eight million, or by more than 57 per cent. But the population of the subordinate urban settlements increased more than five times during the same period. As Figure 2 shows there are many cities in Moscow oblast which have grown at singularly rapid rates. Two cities lying outside the oblast border, but certainly within the Moscow metropolitan area in the opinion of most authorities, have also experienced rapid growth between 1959 and 1981 (cf. Aleksandrov and Obninsk, Figure 2) (Hausladen, 1983, pp. 218-49). One of the objectives of urban planning in the Moscow metropolitan area has been the containment of sprawl and the protection of the forest park belt as a recreation zone for Muscovites. But it is precisely in this intended green belt where the urban population has increased fastest in both relative and absolute terms

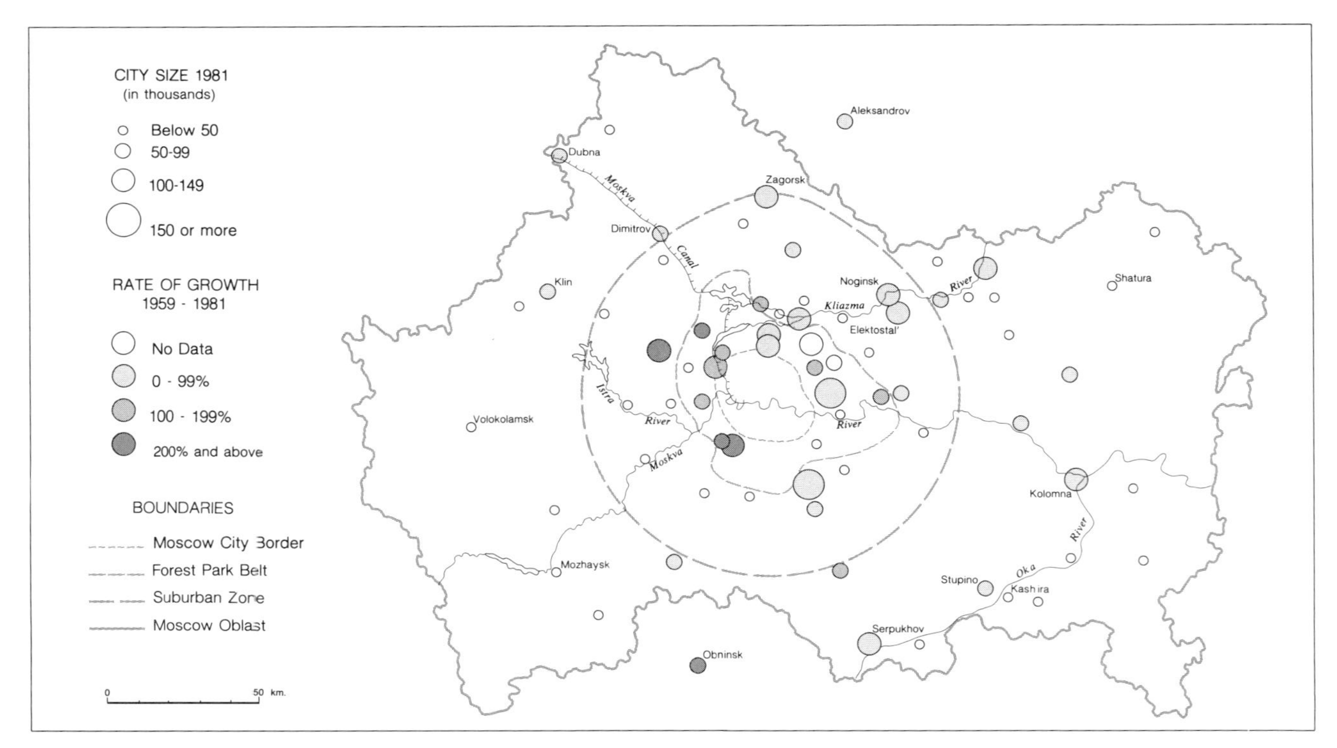

Figure 2: The Moscow Agglomeration

between 1959 and 1981 (Figure 2). The obvious question is, if this can occur in the best planned city-region in the country, what is happening in the majority of cases where planning authority does not extend past city borders?

The vast majority of the Soviet urban population now lives within an agglomeration. The definition of an agglomeration varies, but the term customarily refers to a network of urban and rural settlements linked to a city of at least 250,000 inhabitants by a transportation system that permits journeys from dependent settlement to the central-city within a two-hour travel time.[9] The concentration of population in central cities within agglomerations is higher than in western countries. Such concentration in the central city, which rarely is less than one half of the total metropolitan population of the agglomeration, is at least partly due to the low level of private automobile ownership and to the preponderance of apartments as opposed to detached, individually owned housing (Rowland, 1986, p. 645). Clearly, with such spatial concentration of the metropolitan population the small and remote towns are even further removed from the mainstream of Soviet economic development. The problems posed by this reality are currently being addressed in creative fashion in terms of concepts and theories if not by planning regulation.

The planned development of the urban system has been part of Soviet theoretical formulations from the outset. But aside from a few general locational guidelines there has been a notable absence of specific policies and programs to achieve such an objective. New towns were not to be created by ministries in isolation from the regional settlement system of which they should comprise an integral part, an objective which found legal expression in 1933 as noted earlier. They were to be built according to Soviet planning principles and provisioned accordingly. While more than 1,200 new cities have appeared on the Soviet map since 1917, many in fact suffer from inadequate attention to planning principles. With the demise of the concept of an optimal size city as a useful planning tool in the early 1960s, other strategies gained prominence. A much more overt systems perspective began to evolve. One manifestation was the satellite city program of the early 1960s.

Conceived as a tool to divert growth from the major urban centres, the satellite city clearly demanded a systems perspective if only at the scale of the agglomeration. Most often linked to the writings of the Soviet geographer, V.G. Davidovich, the satellite city

---

[9] For a general discussion, see Lappo, and Pivovarov, 1984.

experiment was endorsed by Nikita Khrushchev and played an important role in the efforts to stem the growth of Moscow during the 1960s and early 1970s. The new town of Zelenograd in the Moscow metropolitan area (formerly called Kryukovo) was initially hailed as the first model satellite city. Under the direct jurisdiction of the Moscow City Soviet, or municipal government, it was for a time in a unique administrative position. For a variety of reasons the Zelenograd experiment did not take root and spread.[10] While there continues to be great concern expressed over the growth of the large city, the data presented in Table 2 underscore the fact that such concern has not had much effect in stemming the trend.

The optimal size and satellite city concepts have both passed from centre stage in the planning and geographical literature. The real change signalled by the satellite city experiment, that is, an explicit recognition of the need to theorise and plan from an urban systems perspective, has gained momentum in the past quarter century however. There are now several research groups formulating plans to guide the evolution of the Soviet urban system. They include a group associated with G.M. Lappo at the Institute of Geography, Academy of Sciences, in Moscow. The effort here is focussed on identifying urban agglomerations, of which more than 70 have been defined, and lobbying to have the urban agglomeration accorded some sort of juridical status, as yet an unrealised objective. The work of this group is derived from, and continues, the earlier research of Davidovich on satellite cities. A second group of geographers has articulated what is labelled a theory of a unified system of settlement. Associated primarily with B.S. Khorev of Moscow State University, this approach explicitly addresses the continuing need to remove the imbalance in development, and hence in the standard of living, between urban and rural areas. Khorev has argued the merits of a scheme that is hierarchical in a Christallerian sense, but explicitly includes the planned integration of the rural sector into the system. A rough equivalent of the urban corridor concept has been developed in the Soviet context by O.R. Kudryavtsev of the Central Urban Planning Institute in Moscow. In common with the foregoing concepts, Kudryavtsev envisions the major urban centres being the locus of future economic growth. The central task remains that of directing the spread effects of development into the appropriate smaller urban centres and rural hinterland.

Perhaps symptomatic of the post-1960s preoccupation with creating a logical framework for regulating the urban system is the plan put forward under the aegis of the Central Urban Planning Institute in the mid-1970s. The so-called General Scheme for a System of Settlement incorporates a number of the basic ideas associated with Lappo, Khorev and Kudryavtsev. It was to address a number of problems associated with a settlement system which had evolved under conditions where the planning emphasis, indeed, authority, ended at the city border. These included, uncoordinated urban growth by city and by region, a lack of economic diversity in many cities, and a lack of integration of urban and rural sectors with resultant discrepancies in standards of living at both the national and regional scales. Introduced with considerable fanfare, the General Scheme was to provide the necessary spatial frame of reference for the development of the national economy between 1976 and 1990. Comprising a network of hierarchically organised settlement systems, 60 large, 169 medium and 323 small, when complete it was intended to embrace more than 90 per cent of the total Soviet population (Figure 3). While in practical terms the General Scheme does not seem to have been of great practical significance since it was never adopted by government as official policy, it nonetheless gives tangible expression to the growing preoccupation of many Soviet analysts over the long-standing problem of how to eradicate the differences between town and country under socialism. Moreover, it reflects implicitly the growing attention paid to the role of agglomerations, perhaps the key feature of the contemporary urban system.[11] Like the other approaches mentioned, it treats the Soviet urban system as a regionally differentiated one. The question remains, however, in light of past practice and present principles, what does the Gorbachev era portend for the development of the Soviet urban system?

---

[11] An elaboration of these schemes may be found in Shabad, 1986 and Portyanskiy, Ronkin, Shchukin, 1986.

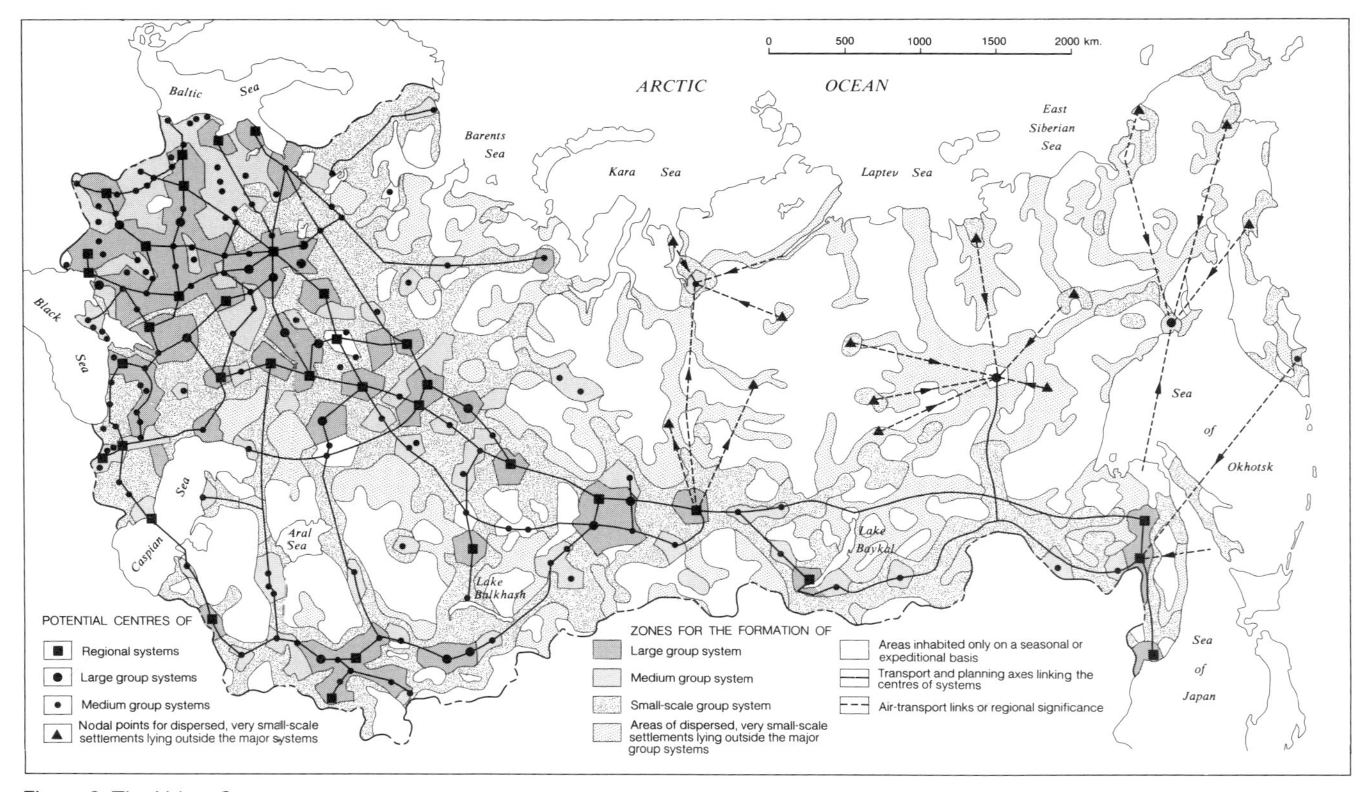

Figure 3: The Urban System

URBAN GROWTH IN THE GORBACHEV ERA

It is not possible to do more than speculate as to the course of urban development during the next few years. However, as a number of trends are already well entrenched the question is whether the arrival of Gorbachev suggests any fundamental change in goals and strategies which might impinge on the Soviet city and the urban system of which it is a part.

The death of K. Chernenko in March 1985 brought Mikhail Gorbachev to the fore as both chairman of the POLITBURO, the key decision-making body in the Soviet Union, and of the Secretariat of the Communist Party Central Committee. In speech after speech Gorbachev has made the relationship between improvement of the material conditions of the Soviet population and the performance of the Soviet economy his central theme. The following excerpt from a May 1985 report on the upcoming 27th Communist Party Congress is typical:

> The development of Soviet society will be determined, to a decisive extent, by qualitative changes in the economy, by its switch onto the tracks of intensive growth and by an all-out increase in efficiency.[12]

At the Party Congress itself in early 1986 Gorbachev continued to espouse these same points. The essential reason for this emphasis is simple - the annual rate of economic growth has declined steadily since the Second World War. For example, since 1966 the average annual rate of growth of national income by five year planning period has slipped from close to seven per cent to less than four.[13]

Clearly, only so much increased production can be squeezed out of any system through appeal to public conscience, and only so much can be done to trim the bureaucracy. And more draconian measures such as those initiated by Yuri Andropov who preceded Chernenko against "idlers" and "parasites" rarely have a lasting effect. Indeed, the initial gains in labour productivity which the Andropov campaign generated have proven temporary and economic growth thus continues to slow down.

----

[12] "Party Plenum looks to 27th Congress," *The Current Digest of the Soviet Press*, 37(17) May 22, 1985, p. 3.
[13] For a general discussion of related matters, see Rumer, 1986.

The problem is not simply attributable to the attitude of workers, but embraces such contentious issues as the style of management and long-standing modes of investment. For example, for years the preferred route to higher production levels was through investment in new plant. Reconstruction, modernisation and maintenance was a distinctly lower priority amongst managers. Past failure to technically re-equip and update existing enterprises has caught up with planners and managers alike. The task now is to find ways of substantially augmenting investment in reconstruction of existing plants without jeopardising development plans requiring new facilities, especially those of a high technology kind. *Perestroyka*, or restructuring, is a Gorbachev era slogan. As most existing industry is, as noted already, located in the major cities, *perestroyka* has obvious implications for the urban scene.

There has long been recognition of the fact that, within limits, there is higher labour productivity and better return on investment in larger cities or agglomerations than in small, remote locations in the Soviet Union. But for ideological reasons such a rationale for development strategy was customarily rejected. Had there not been such rejection much of the economic development, and the growth of the urban system, in the Soviet north and eastern regions simply would not have taken place. Indeed, there appears to be an ideological basis to the current criticism of GOSPLAN, the state planning agency, by some members of the POLITBURO for not moving with greater alacrity in drafting a development strategy for further investment in, and development of, the eastern and northern regions. It is precisely in the east and north where labour turnover rates are exceedingly high, where labour productivity is comparatively low, and where return on investment is therefore less than might be otherwise obtained. As in other situations, the signals from the POLITBURO are confusing. Development of these regions is not consistent with the incessant call for greater economy in factor usage and higher return on investment of increasingly scarce capital. The apparent reluctance of GOSPLAN to proceed with great haste in further developing the high cost eastern and northern realm is therefore quite understandable. While the argument in favour of greater investment in existing large centres because of the external economics thereby gained is not new, it is heard more and more often of late (e.g. Portyanskiy, Ronkin, Shchukin, 1986, pp. 152-53).

Development during the current Five Year Plan (1986-1990) will inevitably have to reflect the pressure for greater efficiency. The past pattern of ministerial investment in basic industry, in research and development centres, in the knowledge industry in general, all of which showed a large city bias, is therefore not likely to be altered. Indeed, it is likely to be intensified. Thus, the trend toward a growing share of the urban population residing in the largest city-size group will continue. Since the integration of physical planning with economic planning as represented by the GOSTROY and GOSPLAN central agencies, has not yet occurred, the prospect for a balanced development of such places and for stemming further development of agglomerations is not promising. The management of agglomerations is exacerbated by virtue of the absence of any widespread network of metropolitan-based planning agencies. Save for a few of the largest cities, such as Moscow and Leningrad, the authority of the City Soviet and its planning bureaucracy ends at the city boundary. Furthermore, the emergence, especially under Gorbachev, of the branch plant phenomenon, and the apparent emphasis on decentralised decision-making, reinforce the existing tendency for managers to search out locations near to, if not in, the country's major urban centres. Small wonder that more than three-quarters of new city formation is occurring within existing agglomerations.[14]

Moscow, the nation's best planned city, was to have had no more than five million inhabitants according to the 1935 Plan. It has more than eight million at present. Without existing controls over migration, investment and so on, the total no doubt would have been greater still. The basic control mechanisms are unchanged. The pressures bearing on location in the large city and its metropolitan area at the present time are probably greater than ever before. Thus, it is likely that the Gorbachev era will also be a period in which the link between principle and practice, between abstract ideal and reality, remains tenuous.

SUMMARY

Since the 1960s there has emerged a concerted effort to turn the attention of all planners away from the individual city to the city region and to the national settlement system. The General Scheme is one manifestation of this tendency, the growing interest in the role of the urban agglomeration is another. Soviet town planners traditionally have had jurisdiction only within the borders of the city. When their policies have been successfully enforced, outlying settlements have been the recipients of much new, and unplanned, investment. New investment produces jobs, and jobs urban growth. Very frequently the places selected for such investment are part of existing, or nascent, urban agglomerations.

----

[14] "Novye goroda v sistemakh gorodskikh aglomeratsiy SSSR," *Referativnyy Zhurnal*, *Geografiya*, *07E*, No. 1, 1985, 1E45, p. 10.

There are many places which do not enjoy easy accessibility to any major Soviet city. Such settlements bear witness to lost opportunities. In the current decision-making environment, which emphasises cost efficiency over spatial or regional equity, the small city remotely located often has a sizeable level of under-employment. But the lack of adequate municipal services, the absence of centres of higher and technical education, and a general dearth of consumer and cultural services, which follow from years of limited investment, are more than sufficient to outweigh any possible advantage of manpower availability to ministerial decision makers. In such communities absolute population decline is not unheard of. Since only a handful of urban agglomerations have been set up as independent planning administrations, since there is as yet no operational plan for managing the development of the settlement system, throughout most of the country the large city grows faster than intended spawning development in its shadow. The small town outside these burgeoning agglomerations is left to get on as best it can with limited investment and resources.

REFERENCES

Andrusz, G.D. (1984) *Housing and Urban Development in the USSR*, London: MacMillan.

Bater, J.H. (1980) *The Soviet City: Ideal and Reality*, London: Edward Arnold.

Bater, J.H. (1984) "The Soviet city: continuity and change in privilege and place," in Agnew, J., Mercer, J., Sopher, D., (eds.), *The City in Cultural Context*, Boston: Allen and Unwin, pp. 134-62.

Demko, G.J. and Fuchs, R.J. (eds.) (1984) *Geographical Studies on the Soviet Union: Essays in Honor of Chauncy D. Harris*, Chicago: University of Chicago, Department of Geography Research Paper, No. 211.

Dienes, L. (1984) "The development of Siberia, regional priorities and economic strategy," in Demko and Fuchs, *Geographical Studies on the Soviet Union*, 189-213.

Fuchs, R.J. and Demko, G.J. (1983) "Mobility and settlement system integration in the USSR," *Soviet Geography: Review and Translation*, 24(10): 547-59.

Harris, C.D. (1970) *Cities of the Soviet Union*, Chicago: Rand McNally.

Hausladen, G. (1983) *Regulating Urban Growth in the USSR: The Role of Satellite Cities in Soviet Urban Development*, Syracuse University, unpublished Doctoral Dissertation.

Hausladen, G. (1984) "The satellite city in Soviet urban development," *Soviet Geography*, 25(4): 229-47.
Jensen, R.G. (1984) "The anti-metropolitan syndrome in Soviet urban policy," in Demko and Fuchs, *Geographical Studies on the Soviet Union*, pp. 71-92.
Khodzhayev, D.G. (1986) "Gosudarstvennoye ekonomicheskoye i sotsial'noye planirovaniye i upravleniye razvitiyem rasseleniya," in *Dinamika Rasseleniya v SSSR. Voprosy Geografii, 129*, Moscow: Mysl', 12-24.
Khorev, B.S. (1986) "Aktual'nye problemy politiki rasseleniya v SSSR," in *Dinamika Rasseleniya v SSSR. Voprosy Geografii, 129*, Moscow: Mysl', 24-40.
Lappo, G.M. and Pivovarov, Yu. L. (1984) "Settlement in the USSR," in Bourne, L.S., Sinclair, R., Dziewonski, K. (eds.), *Urbanization and Settlement Systems: International Perspectives*, Oxford: Oxford University Press, pp. 335-55.
Lewis, R.A. and Rowland, R.H. (1979) *Population Redistribution in the USSR: Its Impact on Society, 1897-1977*, New York: Praeger.
Listengurt, F.M. (1985) "Puti sovershenstvovaniya rasseleniya na territorii SSSR," *Izvestiya Akademii Nauk SSSR. Seriya Geograficheskaya*, No. 2: 68-76.
Loeber, D.A. (1984) "Limitchiki: on the legal status of migrant workers in large Soviet cities," *Soviet Union/Union Sovietique*, 11(3): 303-08.
Narodnoye Khozyaystvo SSSR v 1985 g, Moscow: Finansy i Statistiki, (1986).
"Novye goroda v sistemakh gorodskikh aglomeratsiy SSSR," *Referativnyy Zhurnal, Geografiya, 07E*, No. 1, 1985, 1E45, 10.
"Party Plenum looks to 27th congress," *The Current Digest of the Soviet Press*, 30(17) May 22, 1985, 3.
Portyanskiy, G.S., Ronkin M., Shchukin, Yu. M. (1986) "Problemy razvitiya i modelirovaniya gorodskikh sistem," *Modelirovaniye Geosistem. Voprosy Geografii, 127*, Moscow: Mysl', 146-60.
Rowland, R.H. (1980) "Declining and stagnant towns of the USSR," *Soviet Geography: Review and Translation*, 21(4): 195-218.
Rowland, R.H. (1986). "Changes in the metropolitan and large city populations of the USSR," *Soviet Geography*, 27(9): 638-58.
Rumer, B. (1986) "Realities of Gorbachev's economic program," *Problems of Communism*, (May-June): 20-31.
Shabad, T. (1984) "The urban-administrative system of the USSR and its relevance to geographic research," in Demko, G.J. and Fuchs, R.J. (eds.), *Geographical Studies on the Soviet Union*, pp. 93-108.

Shabad, T. (1986) "Urban issues in the Soviet Union," in Conzen, M., (ed.), *World Patterns of Modern Urban Change*, Chicago: University of Chicago, Department of Geography Research Paper, No. 217-218: 101-28.

Ann Sheehy, A. (1985) "Population trends in the Union Republics, 1979-1984," *Radio Liberty Research Bulletin*, Vol. 29, No. 22 (3331), (RL 166/85): 1-7.

Starr, S.F. (1978) "Visionary town planning during the cultural revolution," in Fitzpatrick, S. (ed.), *Cultural Revolution in Russia, 1928-1931*, Bloomington: Indiana University Press, pp. 207-40.

"The riots in Alma-Ata: what caused them?" *The Current Digest of the Soviet Press*, 39(1) February 4, 1987: 4-6.

Zaslavsky, V. and Luryi, Y. (1979) "The passport system in the USSR and changes in Soviet society," *Soviet Union/Union Sovietique*, 6(2): 137-53.

APPENDIX A

THE MAKING OF A HUMANIST GEOGRAPHER: A CIRCUITOUS JOURNEY

Peter H. Nash
School of Architecture
Department of Geography
and
School of Urban and Regional Planning
Faculty of Environmental Studies
University of Waterloo
Waterloo, Ontario

"The art of progress is to preserve order amid change and to preserve change amid order," said Alfred North Whitehead, Harvard philosopher and one of the most eloquent humanists of this century. We write for progress - our own as well as that of others. It is with major trepidation and minor reluctance that I accede to Leonard Guelke's request to write an essay on the "making" of a humanist geographer, especially because the very nature of the task is autobiographical. It must depict selectively salient encounters on a circuitous journey. Thus it describes forty years of wandering through parched landscapes of geography and itemizes the refreshing humanistic oases experienced on the way, with an audacious look at the promising land of geography in saner times to come. But, primarily, the essence of the essay constitutes what Whitehead so clearly phrased. There appears to be an order in the changes observed, and the mechanisms for change must be preserved in our current order. My "lived experience" leads me to much optimism and I believe it is possible to share it somewhat in these pages. Consider the triple asterisks to be "circuit breakers."

* * *

Curiosity, coupled with imagination, is the embryo of geographical inquisitiveness. The earlier the better. If Lowenthal's statement that "anyone who inspects the world around him is in some measure a geographer," I certainly started my inspections at a very early age from the balcony of our Frankfurt apartment on the Untermainkai. Watching the tugs and pleasure boats on the Main River conjured questions and images in my mind. The same thoughts came to fore during my clandestine sprees to the Hauptbahnhof five blocks away. Where were all these places that these steaming monsters came from and where were they going? Soon my father's huge atlas was my favourite book and I drew my own little maps - some of imaginary places. Many leisure hours were spent sitting on the carpet in his study examining in detail the wonders of his majestic globe. The stories of Karl May, probably about a dozen volumes, such as *Winnetou*, the fascinating accounts of Heinrich Schliemann's excavations, particularly of Troy, and the adventurous expeditions of Sven Hedin expanded the horizon. It is little wonder that my early geographic education took roots. Teutonic *Heimatkunde* in a German elementary school and a more refined facts memorization approach in English high school became a workable amalgam. Geography teachers liked me: I knew my facts and always drew nice little maps and charts to illustrate my answers. It is not surprising then, that when I had to declare a major upon entering university, I selected "Earth Sciences" and chose more geography than geology courses. But even during my undergraduate days at the University of California at Los Angeles (UCLA) I became somewhat dissatisfied with the "scientificness" and academic narrowness with which matters of earlier delight were presented. Only the adventurous lectures of George McCutcheon McBride on his travels in western South America and the illuminating approach of Joseph E. Spencer on the Far East brought satisfaction to a slightly disenchanted youngster.

However, during that period, I had an important outside influence - the first of the humanistic impacts. I became acquainted with Hans Reichenbach in his course on Inductive Logic and decided to continue my philosophical studies with him in his "Philosophy of Nature" course. I expected to be taught the various approaches to studying nature. Instead it was an introduction to "epistemology:" how do we know what we know and ways of knowing, and, to him, epistemology *was* logical positivism. Reichenbach had been a member of the "Vienna Circle," and soon I became thoroughly indoctrinated. My heart beat faster when I heard about the world of macro-dimensions and of the micro-world. I realized that the world of the geographer, at least the way it was interpreted by the exemplars, was the world of the *middle* dimensions. Micro-space and macro-space were based on different axioms. As an obvious outgrowth I also began to think about geographies of the mind, the role of imagination, the impact of the humanities, and even the geography of the future. In retrospect, in my most brazen positivistic exposures I developed some of my most fertile non-scientific thoughts. The relativity theories of Einstein, Heisenberg's principle of indeterminacy, Schrödinger's functions, Haeckel's laws - these were really the exciting ideas that fell upon me during my undergraduate years, rather than Köppen's detailed climatic classification and the memorization of rock characteristics and soil categories. Reichenbach liked my term paper on "Causal Laws and Statistical Laws" and invited me to take his "Probability Theory" seminar. He was revising his *Wahrscheinlichkeitslehre* (1934) at that time, which was ultimately published as *The Theory of Probability* by the University of California Press in 1949. We also used his *Experience and Prediction*, an analysis of the foundations and structure of knowledge, which had been published by the University of Chicago in 1938. I intuitively understood the relationship to geography of such concepts as the truth theory of meaning, the verifiability theory of meaning, projection as the relation between physical things and impressions, the different kinds of existence, and transition from immediately observed things to reports. Reichenbach was also putting final touches on his manuscript of *Philosophical Foundations of Quantum Mechanics* at that time, and I felt honoured that he asked me to help him proofread it. We did have some good conversations about how geography might relate to the disturbance of the object by the observation, the indeterminacy in the observational language, and the limitation of measurability. I also believe that my interest in patterns and cycles in my search for order was stimulated by Reichenbach via the presentation of the works of Joseph Fourier, who showed that even the most complicated forms of vibration of a violin string can be seen as the sum of underlying simpler Pythagorean vibrations. I was too young, too inexperienced, too timid and really too ignorant to attempt any "bridge building" between my philosophical thoughts and geographical factual knowledge, but it made me more responsive to such concepts as George Zipf's "Principle of Least Effort" and Edward Wilson's "Sociobiology" years later.

The cut-off point in this episode came in the final semester of my senior year, and I enrolled once more in a Reichenbach seminar on "Philosophy of Space and Time." I expected this course would really "blow" my mind. Instead, the few students in the class were mostly physicists and mathematicians, the text was Reichenbach's *Philosophie der Raumzeitlehre*, and the classes consisted of discussions of these German reading assignments. I had very serious trouble with the mathematics and considerable difficulty with the technical German. Half way through the semester I felt that failure was imminent, although I am sure this kindly gentleman would not have "flunked" me. But I dropped the subject, my philosophical escapades, my epistemological forays - and for a long time! It was now "Pearl Harbor" time, and joining the Army seemed of much greater importance. During a brief stint as a graduate teaching assistant just prior to enlistment, Joseph Spencer pressed the first installment of "The Nature of Geography" into my hand, which had just appeared in the *Annals of the Association of American Geographers*. I devoured it. I had hoped something of that type, focusing on the philosophy of geography, might exist. And now it had appeared! Some day, I thought, I would like to meet Richard Hartshorne. And, perhaps, after the war, God willing, I might study under him and get my scatterbrained thoughts in order about this discipline with which I had become enamoured.

* * *

In the United States 160th Engineer Combat Battalion and the 12th Army Group Intelligence Service I had my first taste of wonderful practical uses of geography, and I bathed in the glory of my map reading and terrain analysis capabilities. It felt good to be respected not for what you knew, but for what you could actually do to help your comrades to win battles. I was in demand, not so much for my ability to speak French and German, but to assist in tactics and strategies. Geography, I realized, was broadly respected and even sporadically understood by Army Engineers. During combat one day, I noticed that one of my commanding officers was reading Roderick Peattie's *Geography and Human Destiny*. It had been published as a paperback for the troops. The subject matter provided many hours of animated conversation between us and was the beginning of a long friendship.

The climax of my involvement with geography occurred after the cessation of hostilities, when serving as a "Special Agent" with the Mobile Field Interrogation Unit #4. I was asked by Third Army Headquarters to interview and report on General Karl Haushofer. I knew about his works from Jan Broek's "Political Geography" course. Here I was face-to-face with "the" master geopolitician. The confrontation with him stirred feelings of awe and pity. And I was amazed that the General really did not understand the difference between geopolitics and political geography, which Jan Broek had no trouble getting across to undergraduates. He did not comprehend, or perhaps did not wish to, how his geographic thinking had become politicized and warped. Yet he muttered several times a day: "Patriotismus und Religion sind die grössten Feinde des Friedens," a trite homily at the time, but increasingly meaningful to me in this nuclear age of terrorism and self-righteousness. But Hess's geography mentor, and indirect influence on Hitler when Rudolf and Adolf shared a prison cell, had carried Ratzelian thoughts to the extreme so that the consideration of space as an organism was considered axiomatic. I remember wishing that I had Bertrand Russell or Hans Reichenbach with me so that we could shift to a different epistemological mode. But there was a half century of experience separating us and I knew Haushofer had no inclination to question his vulnerable premises. I also felt that he was mendacious and did not wish to acknowledge the extent to which his views were blinkered. Anyway, his "Institut für Geopolitik" had been a minor paper organization. I reported my findings in a "secret" memorandum to Third Army Headquarters. It was published a year later somewhat embellished and without authorization by Father Edmund A. Walsh in a *Life* magazine article entitled "The Mystery of Haushofer" (September 16, 1946). Haushofer and his wife committed a double suicide a few weeks after my visit. I had tried to cheer up Haushofer as much as I could, but it was obvious that he was deeply despondent.

* * *

My next intellectually stimulating experience took place in Grenoble, when on temporary duty at the U.S. Army TWCA (Training Within Civilian Agencies) Center and working for SHAEF (Supreme Headquarters Allied Expeditionary Force) Military Headquarters. It not only included teaching at the University of Grenoble, but also involved the writing of a detailed report on French resistance activities and the German devastations of the Vercors Massif in the French Alps. There I met Raoul Blanchard, the last protegé of Paul Vidal de la Blache. There was an immediate "chemistry" between us - one of the many fortunate occasions in life when I was unexpectedly, but strongly influenced by a magnetic scholar who broadened my horizons and increased my desire to learn by constantly bombarding me with pregnant questions. I confided in him that I was somewhat disenchanted with the narrowness of the discipline of geography. I related to him that I felt that I wanted to do something more useful, and described to him my feelings about the areal devastations I had seen and my thoughts about how the destroyed cities and towns might be rebuilt. Would City or Regional Planning not be a better activity for the post - World War II world? He was wise enough to stress in the beginning of our relationship, not his immense knowledge of physiography, but the fact that he was the founder of urban geography. He presented me with his early studies of Grenoble, Annecy and Nice. The major point he made was that a good urban geographer would not only collect data, but that he would focus on intangibles, such as values, mores and beliefs. In other words, according to Blanchard, urban studies could open the humanistic door to geographers, and they could give him even the opportunity to contemplate the future of such a growth. They could give directions for planners and planning. And he cited many examples. As we tramped through the Vercors Massif and other parts of the French Alps and surveyed the destroyed communities, such as Vassieux and La Chapelle, he made salient comments about the future land use, circulation patterns and other directions these communities might take, based on the facts we had collected. I did not know at that time, or perhaps I paid no attention to this fact, that Blanchard was the first role model of francophone Canadian geographers and there was a time when about half of all the lecturers in France and Quebec had been taught by him. Today he is still one of the best exemplars of first-rate microgeographic field work. In some areas described in his twelve volume work on the French Alps he devoted a page per square kilometer or for twelve inhabitants.

The deepest impression on me was that he stated repeatedly in a variety of ways that he had "*no* philosophy" and took pride in his eschewal of "acral" research. He used this word "acral" in a condescending manner. Webster's International Dictionary actually defines it as "end, extremity, highest point;" also "of or belonging to the extremities or peripheral body parts." I am sure, that when he spoke of "Acral Geography" in a cynical manner, he did not refer to the "highest" forms of the geographer's art, but to those misguided souls who dabbled in the periphery and scholars concerned with the extremities rather than the heart of geography. One of his often repeated dicta is "Some geographers spend their lives trying to define geography. As for myself, I make it!" But I had absorbed enough epistemological knowledge from Reichenbach to realize that he was what would now be called a logical positivist, because he believed that theory is axiomatized when it possesses a set of definitions from which assertions can be deduced as consequences. Development of theory in both human and physical geography were intimately linked. His values did not influence the interpretation of his data; at least he tried to keep his values separate from his facts. To some extent he was also a "causationist," because he had an objective view of man in space and time, with considerable stress on causal concatenations and mutual relationships. He realized that environmental determinism can be applied towards the solution of large practical problems. His work in the Alps had developed in him a special kind of objectivity, which may be described as a religiously based humility. Most of his research was idiographic, and thus the conclusions were non-deterministic and non-mechanistic. Nevertheless his search for laws implies also a nomothetic approach. The causationist approach made him explore environmental limits in which human choice takes place. Often one can see Blanchard's "scientific method" combined with a "humanistic" geography. That perspective brings Blanchard close to being a "realist." This is particularly apt when one sees the naîve realist as a "commonsense" geographer. He probably would agree that human science is an empirically based rational enterprise, and that the critical realist can be a logical positivist. Blanchard had a willingness to open the back door to the humanities via urban geography, but this in no way would influence Blanchard's total concept of his field. He had a clear notion of who *he* was and what *he* wanted. He was not a conformist and his philosophical beliefs were implicit. Little did I think in 1945 that I would re-visit the Vercors forty years later and devote time to studying Blanchard's half-century of productive life and its impact.

* * *

Civilian life returned me to my UCLA graduate assistantship, where I embarked immediately on my M.A. Thesis: *The Vercors Massif: A Geographical Analysis of a French Alpine Region*, dedicated to Blanchard. My advisor was Ruth E. Baugh, who was the last student of Ellen Churchill Semple and completed her final book, *The Geography of the Mediterranean*, during Semple's last illness. Baugh almost worshipped Semple, and much of this admiration rubbed off on me. I was nudged into a more precise style of geographic writing and drifted, without being aware of it, into a Ratzelian mode. But Baugh opened my eyes to the beauty of Semple's writings, which had such a strong humanistic base. Not only the poetry and imagination of her language, but the depth of her knowledge of the classics brought a new dimension to my thinking. Here was a scholar who was also a competent linguist, equally at home in a variety of living and dead languages, including Latin, Greek, Hebrew and Arabic, as can be seen in the fireworks of sparkling footnotes on almost every page of her books and articles. Some serene sections, such as the opening paragraph of "Ancient Mediterranean Pleasure Gardens," have been a delight to me throughout my life. Perhaps I had become a modified environmental determinist via Semple and Baugh during this first year of graduate study. My thesis and first couple of articles reflect this. I might label it as a very mild form of "causationism" now. But through Baugh I developed more profound scholarship by delving deeper into the classics. The only other geographer prior to that, who stressed the classics in his lectures, was Carl Sauer. In her devotion to Semple, Ruth Baugh made me aware of the salient role of the mentor as model to the student. I also realized that, as a female, Baugh was not treated as an equal by her fellow faculty members, but she never once complained about the various types of discrimination she endured, such as not being promoted to a full professor in spite of a high scholarly output. I was therefore delighted when, more than a decade later, after her retirement from UCLA, she accepted my offer to teach for a year at the University of Cincinnati as a Visiting Professor of Geography when I headed the department there.

Now I was ready to leave Southern California for the heartland of U.S. geography. Madison, Wisconsin was my choice. I had written Hartshorne a couple of letters and I was offered a graduate instructorship. Finch and Trewartha were there - authors of the most widely used college introductory geography textbook. The department was run like the army, and

> General Trewartha was in command. If you knew your climatic facts you were a good soldier, and as a drill sergeant I had to drill them into the students. The Finch and Trewartha text was considered the "bible" and graduate seminars were recitation sessions. Students were evaluated on the basis of the amount of facts they had internalized, regardless of background and level of education. Once I was called into the chairman's office and dressed down for reading to my freshman introductory physical geography class a couple of pages about the cyclone "Maria" from George R. Stewart's *Storm*. "The reading of fiction has no place in a geography class!" I was told. No *Lebensraum* for the humanities here! I realized the hopelessness of my explanation as it produced a sarcastic smile. I had shown Trewartha the frontispiece, which contained this quotation from Sir Napier Shaw's *Manual of Meteorology*:
>
> Every theory of the course of events in nature is necessarily based on some process of simplification of the phenomena and is to some extent therefore a fairy tale. (I, 123)

I was shaken by the complete lack of understanding. For the first time I became aware of the reality that I was in "*Science* Hall" not only in name, but also in spirit. The anatomy laboratories were on the next floor up, and the geology labs on the floor below. Positivism reigned supreme. I was disappointed that Hartshorne was part and parcel of that culture. His "Political Geography" course stressed place names, boundaries, languages and dates and you were examined on these. The "Philosophy of Geography" course consisted of plowing through *The Nature of Geography* from cover to cover. Hartshorne wanted to see whether we had absorbed the book. Disagreement was tolerated as long as you agreed with him towards the end of the discussion. The course was uninspiring, even though the students were first-rate, evidenced by the fact that most in the class became chairmen of geography departments in North America within a few years. My own enthusiasm was severely dampened, even though I had numerous delightful private conversations with Hartshorne. But I felt that his thinking was not tied to the humanities; it had teutonic depth, but it lacked breadth in the social sciences and the humanities. Perhaps I had expected too much of this modest, hard-working and dedicated scholar who had, in the eyes of many, become the "pope" of American geography.

During my year at Madison I took a couple of courses from John M. Gaus, Professor of Political Science, who was a specialist in regional planning. A kind and inspiring teacher, who had served under Franklin Roosevelt in the National Resources Planning Board, exposed me to a new variety of scholarly writings on planning theory and public administration, including his own works. Thoroughly acquainted with geography and several applied geographers, he made me realize that regional planning was a sound extension of my interests, and that it involved more than delimiting watersheds, growing seasons and ecological niches. Under his guidance I read Lilienthal, Hayek, Wootten, and above all, Patrick Geddes, the great biologist/sociologist/planner/architect, who was Gaus's hero. The fire was lit. Gaus was under frequent attack for his support of liberal causes. One day he announced tersely that he was relinquishing his Wisconsin professorship and had accepted an appointment in the Department of Government at Harvard University. Later that week he asked me whether I would like to continue my studies at the Harvard Graduate School of Design (GSD), and stated he would see to it that I would receive either an assistantship with him or a scholarship. He did not have to wait for an answer. Blanchard might have scolded me for getting into an "acral" area, but having just read Lilienthal's *Democracy on the March*, with strong pleas for making science useful and the Tennessee Valley Authority as a model, I was ready to march!

* * *

In order to study City and Regional Planning at Harvard University and to receive the professional master's degree, it was necessary to register in the Graduate School of Design. The greatest impact on me during my first year at Harvard was the immersion into the third dimension. Architecture meant the creation of new landscapes and I discovered the plastic element of design and its importance. If you don't like a certain type of music, you don't have to listen to it. If you don't like a work of art, you don't have to go to the museum. But if you don't like a building, you frequently still have to look at it or even live in it. The products of architects and landscape architects are in full public view, and our products were pitilessly scrutinized. Technical proficiency and social conscience were taught simultaneously and often jointly by Walter Gropius and Martin Wagner. The "Bauhaus" spirit prevailed and everyone was forced to "design." The GSD was a seedbed of original creativity and frenetic activity, and Gropius was the catalyst. His philosophy was based on two ideas, one practical and the other aesthetic. Modern designers should get workshop training and should be made familiar with the materials and machines used in mass production. Gropius' aesthetic notion was also deceptively simple. Recognizing that the beauties of nature's creation are part and parcel of their functions, he argued that man's creations should also combine usefulness with beauty. (Albers, Bayer, Breuer, Feininger, Kandinsky, Klee, and Moholy-Nagy had all worked with him.) Gropius felt that the main reason for architectural ugliness is inertia of the heart and that man still clings to some "visible reminder of Grandpa." Martin Wagner, former Chief City Planner (*Oberstadtbaurat*) of Berlin, was in charge of the planning studio, and translated the Gropius philosophy to a larger scale, but gave it sound engineering and economic injections which were frequently very painful. G. Holmes Perkins, later Dean of the School of Fine Arts at the University of Pennsylvania, administered the program and gave instruction jointly with Edward Ullman. Creativity was unleashed and neighbourhood designs were ruthlessly criticized. A three semester course on the "History of Urban Design" brought depth to my regard for urban geography, which was heightened also by courses taken from Edward Ullman and Walter Firey, and research work with them.

* * *

At the other side of the Harvard Yard, in the "Institute for Geographical Exploration" on Divinity Avenue reigned Derwent Whittlesey, the wisest and most gentlemanly geographer I had encountered. Whittlesey taught a field course on "The Boston Metropolitan Area," taken primarily by architects and planners. He had told me that his interest in cities had been kindled via Colby, and especially Blanchard during the latter's frequent stints at Harvard, particularly while the Grenoble geographer was undertaking his detailed studies of Montreal and other Quebec cities. Whittlesey's geographic studies of cities had a firm historical base, and his useful concept of "sequent occupance" was invariably demonstrated by him in the field. He explained how human occupance of an area, like other biotic phenomena, carries within itself the seed of its own transformation. The concept has been erroneously criticized for a number of reasons, even though one critic shrouds his attack by explaining it, in John K. Wright's terms, as a manifestation of "human nature in geography." The invalidity of the criticisms, the demonstration of which I have published, illustrate the sloppy results when geographers fail to examine the original sources and thus make faulty interpretations. Whittlesey's concept of "compage," a neologism for "region," was also inappropriately attacked. It was developed in our graduate "Seminar on the Region," the bulk of which was devoted to a critical revision of Whittlesey's chapter on "The Regional Concept and the Regional Method" written for publication in *American Geography: Inventory and Prospect*. Unfortunately the term was never broadly accepted, in contradistinction to "sequent occupance," and seminars still waste time discussing the geographical meaning of "region." But beyond being an original thinker, Whittlesey's strength was that he could link the perceived past with the envisioned future. His urban studies had four elements: city functions, urban forms, locus, and site. He showed how the forms express the functions, always in terms of the locus and the site. He was hurt that some colleagues felt that he accepted uncritically the Davisian assumption that process is implicit in stage. At the top of a page lying on his desk, entitled "Suggestions for Function of Geography," which I found shortly after his death and have kept as a memento, he had written: "Assumed Purpose: To Study the Present in the Light of the Past for the Sake of the Future."

* * *

At the Graduate School of Design I worked in the Robinson Hall studios almost daily into the middle of the night. My classmates and I found it difficult to concentrate on our drawing boards during those beautiful spring days in the Harvard Yard when the lawns are not yet roped off for Commencement. More than the fragrance of the blossoms, it was the exuberant sound of the Harvard Glee Club/Radcliffe Choral Society that filled the air. It came from Emerson Hall, next door. After a while I could hum along with the Beethoven *Missa Solemnis*, and one night, during the *Gloria*, I could stand it no longer, walked over to Emerson, stood in the back, and had my first glimpse of Serge Koussevitzky conducting in rehearsal.

Omitting a series of events which lead to our friendship and get-togethers at Tanglewood, Massachusetts (the summer home of the Boston Symphony Orchestra), a warm, personal relationship had evolved. Koussevitzky saw in me a devoted young admirer who wanted stimulation, not personal musical advantage. To him I was the honest voice of an appreciative amateur. He liked to talk about geography, cities, and architecture. I thoroughly enjoyed conversations about music. At times we conversed about both concomitantly. He became quite fascinated with some of the concepts I had discussed with Whittlesey, especially the distribution of musical "zones," influences of climates on musical styles, and variations in natural resources as reflected in musical instruments and related compositions. But musical pieces, we found, were also related to architecture, because they have structure and each one is a different composition.

We quickly realized that intellectual activity, whether functional imagery in architecture, geography or music has inevitable emotional consequences. Initial questions always generate complex answers. I believe I realized then, even before I became a professional planner, that emotion should be directed by reasoned thought. Feelings have to be channeled, as Gaus and Gropius had frequently preached. In the process of channeling we explore and discover information outside our fields of specialization. The geographer listens to the *Alpine Symphony* and the musician climbs the Alps. "Apogee" and "gravity" emerges, for example. You can only climb so high and then you must return. When you throw a ball high in the air, it eventually stops and returns. This idea was comprehended so well by J.S. Bach, Walter Christaller and Ebenezer Howard. Reasoned thought has to be absorbed via scrutiny of the work of masters. After an individual "performance," it should be compared to that of a "master." Koussevitzky compared recordings of his own concerts with Toscanini and Bruno Walter; Whittlesey compared his field techniques with Semple and Sauer; Gropius his design methods with Le Corbusier and Gaudi. Koussevitzky taught me that you can never study a classic work enough. You always read new thoughts, hear new sounds, see new sights. "Always gain new insight into works in great detail, even though thoroughly familiar with them" was his prescription. (At my age I have to do it anyhow to refresh my memory!) Performance, whether in the "field," in plastic design, or via musical instruments serves as an essential corrective to analysis as well as an "automatic pilot," because you can develop technical command to a much higher level as you refine your techniques to meet your demands. Also it tops a vein of experience of which analysis *per se* is utterly ignorant. Most analytical procedures have nothing "literally" to do with music, geography, planning or architecture "out there," where the "performer" is. They are interesting from an abstract point of view and misused analytical systems can produce a debilitating involution that blinds one to the performance. The best analysis modifies the product. Owing to this apprehension I placed my interest in the analysis of music vis-à-vis geography in cold storage until I met Max Rudolf, conductor of the Cincinnati Symphony Orchestra, whom I had asked to share a couple of programs on my TV series on "Art and Society" a decade later. Working with Rudolph on a script for a program discussing diffusion of musical styles and areal variations of melodies, performances, preferences, and so on, I harnessed much of what I had thought about in previous years, stimulated by Whittlesey after Boston Symphony concerts and by Koussevitzky at Tanglewood after conducting classes and walks along the lake. This resulted in two lengthy Cincinnati television discussions, with Rudolph at the piano. Some years later, after further developing these thoughts and with appropriate maps, I presented a paper on *Music Regions and Regional Music* at the IGU (International Geographical Union) Congress in New Delhi, where it received much attention in the press, including negative and positive editorials. Some felt that geography was encroaching on areas irrelevant to it; others thought mistakenly that I would be playing recorded music of various countries as background to my lectures. Yet the consensus was that a novel area of abstract thought had been introduced in a practical way. Further refinements were made in *Music and Cultural Geography* and revised in *Music and Environment: An Investigation of Some of the Spatial Aspects of Production, Diffusion, and Consumption of Music* in the Journal of the Canadian Association of University Schools of Music. It has been reprinted in a textbook, entitled *The Sounds of People and Places.* The "Geography of Music" has now become accepted. At the 1985 AAG (Association of American Geographers) Meetings in Detroit three sessions were devoted to that topic. The frontier of geography, if there is such a phenomenon, has been pushed further into the humanities resulting in a better understanding between the most abstract of the fine arts and man's rootedness to the earth.

* * *

A bright star, whose light was suddenly extinguished because of his death, was George Kingsley Zipf, who was a geographer, but totally unaware of it! He had majored at Harvard in mathematics and minored in astronomy, physics and chemistry. His Ph.D. was in Comparative Philology. It occurred to Zipf that it might be fruitful to investigate "speech" as a natural phenomenon, much as an entomologist might study the tropisms of an insect. He wanted to study speech in the manner of the exact sciences, by the direct application of statistical principles to objective speech phenomena. Later, when I got to know him, he had extended his research to other kinds of human behaviour: psychology, economics, advertising and - yes! - geography. His investigations involved statistics for observable empirical social laws, and he attempted to explain and coordinate the laws observed. The ninth chapter of his *Human Behavior and the Principle of Least Effort* was entitled "The Economy of Geography," and it was one of the first publications to discuss Christaller's concepts. Koussevitzky enjoyed his analyses, mathematical though they were, of the works performed by the Boston Symphony Orchestra in relation to repetition versus diversity. However, what impressed me most at that time was his study of the frequency of occurrence of words as well as their distribution. He demonstrated a significant mathematical relationship between the frequency and rank of each word, a relationship that manifests itself equally in Chinese, Gothic, Old High German, Yiddish, Modern Hebrew, Norwegian, American Indian dialects and in numerous English writings. Brief though our encounters were, they were intense. They taught me that a scholar, such as Zipf, could accomplish outstanding work simultaneously in semantics, psychology, sociology and geography, as well as in economics and mathematics. He was in his late forties when he died. Little attention has been paid to his writings ever since.

* * *

One of the strongest influences in my academic career, if not the strongest, was Arthur A. Maass. I had taken a course "Resources Conservation and Government" from this young professor of government. He became a member of the small official "McKay Committee" to report on the future of geography at Harvard, and Edward Ackerman, Henry Kissinger and several others, myself included, met with him on several occasions. Never did I hear "geography" so brutally attacked as a discipline! Maass had read Hartshorne, and Kissinger said that he did. Their consensus was that if the best geography had to offer was this teutonic compilation of facts, a great university under financial stress might be better off without it. It is a pity that other works, representing more "humanistic" scholarship, such as Semple's *Geography of the Mediterranean*, were not selected for reading. Geographers invited to the meeting were rather mute, perhaps fearing an *Asinus asinum fricat* accusation. Isaiah Bowman's report to President Conant was not at all helpful. However, because of these experiences I came to two conclusions. I would be a practicing planner, and I would ask Maass to be my Ph.D. dissertation supervisor. His knowledge of government was vast, but he also had a firm knowledge of geography and planning and understood the relationship between these fields. In seminars, he and Carl Friedrich had already planted the seed for the investigations to be undertaken and my theoretical discussions with Maass developed the framework. Studying his draft of *Muddy Waters*, I realized that it was mandatory that any analysis must be based on some fundamental concepts. I began to make sharp distinctions between "value preferences" and "criteria for evaluation." In his article on "Gauging Administrative Responsibility," Maass actually labels value preferences as "working biases." They are not the criteria through which data or cases are evaluated, but they constitute the basic logical beliefs of the examiner and may be considered quasi-axiomatic with reference to the subsequent analysis. So I had to force myself to make a long list, consisting of summaries of strong beliefs and convictions basic to my analysis, and arranged them in order of scope, starting with the broadest and most general bias. "Criteria for Evaluation," on the other hand, although they may be based on "maxims" referred to above, are largely substantiated by the conclusion of authorities and previous empirical data. I divided these criteria further into "deductive" and "inductive," with the latter group divided into "hortatory" and "heuristic" categories. These were important steps in my theoretical framework, and the philosophical approach provided the "key" to the understanding of my basic hypotheses focusing on the responsibilities and limitations of the professional planner in urban management. From then on the rest was relatively simple, even though the thousand page tome, involving seven case studies, took two years to write.

* * *

After several years of professional planning at the state, federal and local level, I resigned as director of the Planning Department of the City of Medford, Massachusetts. At that time it was a great personal satisfaction to me actually to see a few redeveloped city blocks and major new public buildings in Medford and to be able to point to them, and say to myself: "These structures are here solely owing to Nash's professional efforts!" I realized that a geographer had changed the human landscape of a portion of a city. But the attraction of the calmer academic life became continually stronger, and I accepted the position of Associate Professor of City and Regional Planning at the University of North Carolina. Except for a seminar on Planning Theory, my teaching was of the "nuts and bolts" variety. Education of planners at Chapel Hill was still primarily in the "architecture and engineering" mode, untouched by the humanities. So it was with pleasure that I accepted two years later the headship of the Geography Department at the University of Cincinnati, which had just separated from geology. Soon it was molded into a "Department of Geography and Regional Planning." Sitting in Nevin Fenneman's former office, I thought that the earlier occupant might smile if he saw how the "circumference" of geography had expanded in his beloved building. But the planning portion attracted an increasing larger number of students, and when I left for Rhode Island four years after my arrival, the planning program became an independent unit and an "acral" portion of geography was severed. I regretted having to terminate my part-time studies at the University of Cincinnati Law School, where I had completed all first year courses. Originally stimulated by Charles Haar, the Harvard expert on land use and zoning, I had become increasingly fascinated with the geography of legal systems and had even published a couple of articles on the geography of legislative reapportionment. Yet a return to New England was welcomed, especially by my family.

At the University of Rhode Island, where I became both Dean of the Graduate School and Director of the newly established Graduate Curriculum in Community Planning and Area Development, I had, for the first time, some administrative clout. One of the first steps we took in developing the curriculum was to establish the "Interdisciplinary Seminar" on "Trends in the Contemporary Environment." It was the core of the first year of graduate study. Meetings were held twice a week for three hours throughout the year, limited to the twenty-four entering students. Eight faculty members participated at every meeting, and the clout mentioned made possible the necessary adequate financing and the required regular attendance of all faculty as well as students. The interdisciplinary objective was achieved as each professor chaired discussions based on readings from books selected by him for a period of four weeks. Involved were the fields of architecture, civil engineering, economics, geography, landscape architecture, philosophy, political science and sociology. (The professional planning faculty was required to attend also.) It was gratifying to observe how, over the years, ideas were sharpened and defined within the disciplines by the participating faculty, especially by geographers Lewis Alexander and Edward Higbee. Their publications, although very good prior to that time, subsequently achieved higher quality because of their exposure to the varied humanistic influences. Never before or since have I seen an educational effort yielding such success as that core seminar in Rhode Island in the sixties. But it needed adequate funding and clout, and both vanished from the scene in the seventies. It is very fondly remembered by alumni and faculty.

* * *

My sabbatical year as Visiting Professor in the Institute of Human Sciences of Boston College gave me time to reflect upon my relationship to geography after a long period devoted primarily to more professional activities, including planning, university administration, and consulting. During this time the "Quantitative Revolution" was born, peaked and began to subside. Much of what I did read in the scholarly journals had little appeal or was too mathematical for my old-fashioned taste. But during this period I had another one of these encounters in life with an individual who provides a whole new dimension to understanding and who gives impetus to a fresh approach to problems and issues. The first time I heard Constantinos A. Doxiadis speak was at a banquet in Cincinnati, Ohio. During a "question and answer" period after the presentation of his planning and housing designs, I questioned him, with considerable trepidation, rather severely about some of his assumptions and conclusions. Much to my surprise, during a conversation after the dinner, he invited me for a six o'clock breakfast the following morning. My reluctant acceptance paid off. He requested me to do some designs for him, which lead to invitations to Greece and participation at several "Delos Symposions." Doxiadis, charismatic and peripatetic Greek architect/planner/philosopher and inventor of Ekistics, the science of human settlements, could easily have "passed" as a geographer. He had an uncanny ability to analyze and synthesize, and his "anthropocosmos" model provided a counterpoint to Christaller's "central place" theory. Doxiadis mixed teutonic order and efficiency with Mediterranean flair and artistry. This is not surprising for a Greek who studied architecture in Berlin. To me the Ekistic Logarithmic Scale provided order in my thinking. His propensity to decorate his research with a large assortment of neologisms provided a distinct flavor to his unique style. The elements of the model, especially the "synthesis" category, gave the approach sufficient flexibility for strong humanistic infusions via its vertical dimensions of culture, history, political science and economics. In time, Ekistics became to me a quasi-epistemology at the hard core of a spectrum ranging from realism to existentialism. It has scienticity, if not scientificness, yet leaves ample room for imagination, design and a multiplicity of humanistic adventures. Ekistics developed at the time "Regional Science" was born, and Doxiadis demonstrated much congruence between the two fields, except that ekisticians were almost exclusively concerned with human settlements and were not so quantitatively inclined. I viewed Ekistics in a variety of ways. In a recent article I referred to it as a "syndicated symbol" synonymous with "systematic synchronized synergetic synthesis." From a geographic vantage point, it became almost a romantic icon that linked physical and human geography into a unified conceptual bundle. Jean Gottmann, who was the innovative president of the World Society for Ekistics for a couple of years, seemed to agree with this. And Walter Christaller and Emrys Jones appeared to hold similar views. There was also unanimity in our evaluation of Doxiadis as one of the few original twentieth century synthesizers, not perhaps of the stature of Patrick Geddes, or his disciple Lewis Mumford, but certainly a person endowed with charisma, originality and a deep concern for mankind's past and future.

* * *

It was Gottmann who helped to clarify issues on a panel I arranged on "The Role of Value Judgements and Programming in Geography" at the Association of American Geographers Meetings in Pittsburgh in 1959. Walter Isard had described the geographer as one of many scientists in the broad field of regional science and limited his functions in terms of fact gathering, observation and description. He stated that practical involvement demands specific training in mathematics, especially when dealing with inter-regional social accounts. David Lowenthal, on the other hand, cautioned against research motivated by goals other than curiosity and a zest for discovery, investigation and analysis. To be sure, he said, complete objectivity is unattainable by anyone: the perception, selection, ordering and presentation of data are necessarily subjective and idiosyncratic; geographical thoughts invariably reflect personal and cultural values. He indicated that the scholar who perceived his own biases is best able to approximate a true picture of the world, and urged that values be studied as part of geography as passions and prejudices are crucial determinants of the cultural landscape. Gottmann found a meaningful conciliation between these views by taking a broad humanistic perspective. Since early days, he argued, geographical data have been systematically gathered and classified to help very practical pursuits, such as navigation. Geographers did not always have first hand information on the areas they showed on their maps and charts. For instance, the "predication" of *Terra Australia Incognita* on the old atlases was based on an erroneous idea of the earth's balance. However, Gottmann pointed out, it led in the eighteenth century to the useful discovery of Australia, New Zealand, and the Antartic continent. The fear of being "perhaps wrong" should not deter geographers from forecasting and programming, and predictions "may be useful while wrong too." Geography cannot be truly "experimental," but Gottmann said that our way to experiment is in such deductions, projections and forecasts, and these are always worthwhile when formulated with care and common sense. It was a masterful performance of a humanist geographer, one of the few "renaissance" men in the field, who brought immediate peace to a debate at a time when the quantitative revolution was at its height.

* * *

It was on one of the Delos Symposion Aegean Sea trips that I first encountered some labels that since have become part of my vocabulary. Arnold Toynbee used the concepts of Apollonian and Dionysian influences in an after-lunch conversation. He explained how Dionysian (creative-passionate) pulls and Apollonian (critical-rational) pushes have forced man (or "anthropos" as he called him/her, to avoid being chided by Barbara Ward and/or Margaret Mead) to contemplate who he is and what we might become in relation to alternative environments. Obviously we are mixtures of both. Apollonians are generally balanced in character, rational, nomothetic, temperate, restrained, orderly, measured and striving for harmony. Dionysians are sensuous, irrational, passionate, unbound, lawless, ecstatic and creatively striving. After checking further the writings of Nietzsche, as advised by Toynbee, I found that these labels could easily be applied to geographers. Which colleague would you select to accompany you for geomorphological field work in the High Arctic? Apollonians tend to develop established lines to perfection. And how about a companion to undertake folk music research in the South Pacific? Dionysiac thinking relies on intuition, takes unexpected approaches to research and is frequently subconscious until it nears the end results.

In these discussions the concept of "iconophobia" emerged - the fear of images, a distaste for labeling which some of us deplored. I have overcome this reluctance via the humanities. One set became very meaningful, and I have found it useful not only in the teaching of beginning courses in "Future Studies," but also in the "Nature of Geography" course. The Greek immortals can be guides to learning: the Titan god Prometheus, the "Great Rebel" and father of technology; the capricious sea-god Proteus, the "shapeshifter," who would change his shape to avoid performing his function of foretelling the future; and the life-loving Sisyphus, sometimes called the proletarian of the gods, the "rock roller." Prometheans emphasize faith in science, control, technology, logic and prediction. They believe that personal behaviour can be directed by reforming and controlling the environment. Many applied geographers, city planners and architects fall into that category. Proteans focus on changeability of all life forms, including the human, and on continued transformation through a merging with the environment. Sisypheans recognize the determining effects of the environment, but do not seek to control it. Similarly to the Proteans, they believe that the future depends on inner or personal transformation. Wholeness is achieved within the person, and it is conciliatory and re-conciliatory. It is a process of self-discovery, self-actualization and bringing full potentiality to consciousness. We had all three types represented among the thirty or so participants from most academic fields, professions and business on board ship in every symposion in which I participated. For example, Buckminster Fuller, president of the World Society for Ekistics for two terms, was an ultra-Promethean. Margaret Mead, to me, was a Protean, and Marshall McLuhan was a Sisyphean. Doxiadis himself was aspiring to be a Sisyphean in spite of his Promethean inclinations. There is no doubt in my mind that this classification of ancient origin can also give young geographers clearer visions of themselves as they contemplate their own relationship to the environment.

Another humanist concept used by Toynbee at all the Delos Symposia I attended, which has become increasingly significant to me, is the abstract term "karma." It means the momentum of past action on the present - good and bad. Karma is inescapable, but it is not doom or fate. Karma is created by man's actions and can be re-enforced if good, and mitigated if bad. This "historical process," and this is what Toynbee really meant, helps us to realize not only that man's span of life is short, but that each generation bears the responsibility for the world's affairs for only the few decades of his middle-age. The young, who feel impatient that the middle-aged will have wrecked the world before they take over, do not recognize the force of karma. They too will find that they will suffer from having to "cope with karma:" they will not have a clean slate. To me this concept has not only been an important bridge of understanding between rich and poor, but also a link between the technological and humanistic aspects of our field, between the unchangeable past and the anticipated futures. The present is just a perennially shifting boundary!!

This level of thinking was taken a step further via conversations with Buckminster Fuller on board ship on the Mediterranean and on later occasions, where he expressed his conviction that most of what transpires within human activity is invisible, unsmellable and untouchable. His view was that almost all (99 percent) of "reality" can only be comprehended by man's metaphysical mind, guided by something he only sensed might be the "truth." And he said anthropos *was* "metaphysical mind," and the brain was only a place to store information. The mind communicates, the brain cannot. And so "Bucky" referred to anthropos as a self-contained microcommunicating system and *humanity* as a macrocommunicating system, and concluded that we are only using one percent of our capacities to perceive the "truth." Bucky made it very clear, as did Toynbee, that, until a few decades ago, human beings have been held utterly responsible for their every waking deed. It was assumed that anthropos was consciously controlling all his actions. Freud and Mesmer greatly interrupted this conviction, and proved via hypnotism that human beings were often responding subconsciously to unfamiliar organic and psychological controls. The common criminal laws had to be changed to cope more logically with the subconscious, ergo, unblamable behaviours of anthropos. Fuller explained that physics, chemistry, genetics, virology, physiology, neurology, ecology, geology and even astrophysics are combining to demonstrate that humans are "myriad-frequency complexes of 99.9 percent automated organic processes interacting with the universal environment." So he concluded that, dismaying as it may be to their ego-vanities, human beings are less than one millionth of one percent conscious of all that accounts for their successful growth, adaptation and survival. Thus I realized, even if Bucky exaggerated somewhat, that the human geographer is concerned with that one millionth of one percent conscious control that can be, and sometimes is, amplified by creatively inspired and unselfishly minded scholars. Such individuals are mentioned in this memoir. They make regenerative contributions towards humanity's subconscious success as well as to its conscious satisfaction via the human sciences and the humanities.

* * *

The most recent stage of my intellectual journey started a decade and a half ago when I was asked to head up a new Faculty of Environmental Studies at the University of Waterloo in Ontario. It seemed almost a tailor-made opportunity to be a dean of a faculty that consists of a School of Architecture, a School of Urban and Regional Planning, a Department of Man-Environment Studies and a Department of Geography. Even though it meant moving to another country, the possibilities of this administrative organization, unique in the world, could not be ignored, and I accepted with pleasure. I have not regretted it, even now, when working under the leadership of my second successor a decade after termination of my five years in office.

The boundaries between the units, which I had hoped would become permeable membranes, are still strong. But there are many evidences of continual weakening. The influences of the humanities, though stronger, are counterbalanced by rapid growth of the scientific aspects of environmental studies, which will increase in the future, given the economic conditions and the technical character of the University of Waterloo. But a well-functioning, widely respected and internationally known major academic unit is in place.

Two salient thrusts are shared by all four units in the Faculty of Environmental Studies. One is the emphasis on "ecology," ranging from *Standortstheorie* to ecosystems analysis, which permeates their endeavours. The other is the shared emphasis on the "future." Goals are invariably future-directed, and there appears general consensus that the longer-range aspects must be given more attention. In geography, this concern with application for future benefits had been a major concern of mine since 1960, when I participated at the founding meeting of the IGU Commission on Applied Geography in Stockholm. An international network evolved under the leadership of Omer Tulippe of Belgium, and under Michel Phlipponneau of France after Tulippe's death. My association with them, and such members as Sochava (U.S.S.R.), Shafi (India), Hartke (West Germany), Freeman (England) and Beaujeu-Garnier (France), made me aware of the much broader scope of geography and the need to broaden intellectual horizons. Values, goals, feelings, beliefs, attitudes and visions were the salient issues of these meetings in Prague, Liège, London, Rhode Island, Brittany, Yokohama, Recife and also in Waterloo. We realized how much we were all concerned with studying the future, and that *Gedankenspiele* are a humanistic activity. It was largely the stimulation of many of these encounters that made me channel my activities into "future studies" these past years, including involvement in the founding of the Canadian Association for Future Studies in 1976. Yet the broad exposures to a variety of humanistic influences of respected colleagues from many nations made me question my own axioms and my own ways of knowing. (The battle to make "applied geography" acceptable started in 1960 in Sweden and came to an end in Paris in 1984, when the IGU Commission was dissolved. It is no longer an issue! About one hundred papers were presented at the Eighth Annual "Applied Geography Conference" at North Texas University in 1985).

* * *

The writings of Yi-Fu Tuan had already softened me up for encounters with Anne Buttimer and Leonard Guelke. The charm and persuasiveness of the former and the thorough logical incisiveness of the latter forced me to re-examine many basic tenets of my geographic thinking. The thought-provoking discussions with Guelke of his views on "Idealism" in geography have raised some doubts about my icon of the unity of physical and human geography. His lucid writings liberate many doubts about concepts promulgated by Whittlesey, Ratzel, Semple and even Blanchard. We acknowledge discord in the "church," and I feel at ease to be one of the many "liberation theologians." Even former hard-core conservatives are re-thinking their positions, though not quite the same way as Guelke advocates the re-thinking of the thoughts of occupants of space in earlier times. It is a painful process, especially when the realization dawns that we no longer have a clutch of cardinals and a bunch of bishops dispensing unquestioned dogma. Infallibility is no longer even thought about, and a dozen epistemologies are sprouting. As illustrated so well in the two hour video-taped discussion between Buttimer, Guelke and myself a couple of years ago, accounts of "lived experience" give a vibrant texture to a previously more stolid area of investigation.

A kaleidoscopic vibrance will be evident as more geographers than ever will reflect upon decades of scholarly and professional activities and life-long influences that give meaning to their research. Scientific liberation came first. Some saw the quantitative revolution as a paradigm swing. It peaked rapidly. Non-scientific liberation is much slower and frequently unobtrusive. The slowest intrusions have always been from the humanities. A feeling for them has to be developed. It is not a learning skill. Impacts are generally not sought specifically. They are accidental: humanistic concepts, ideas, values and beliefs find a "prepared" mind, ready to utilize them objectively (and sometimes not so objectively) to satisfy our curiosity about how much of what is where and why it is there. They aid in the comprehension and retention of geographical knowledge. Liberation into an art comes *after* scientific liberation.

My hour-long television interview with Buttimer at the 1984 I.G.U. Commission Meetings on the "History of Geographic Thought" in Geneva was an unexpectedly strong stimulus for self-analysis. I realized that my life-trip had been constantly evolving and was not a deliberate journey. I expect it will thus continue. Geography is a point of departure, but also always a return destination. At best, it is a leisurely trip, with no end prior to the final resting place; it cannot be generalized except via case studies. I have been fortunate that, over four decades, I have had a couple of dozen strong influences.

Almost all were academicians (teachers and colleagues), about half were geographers and many became personal friends. With no Davisian implication intended, five stages can be perceived on the trip. The first might be called *"Awareness of Uncertainty"* (Baugh, Haushofer, McBride, Reichenbach, Spencer), always a healthy sign of youth. With beginning maturity there was the *"Return to the Core"* (Blanchard, Gaus, Hartshorne, Whittlesey), followed by *"Expansion to Application and Design"* involving architecture and law (Doxiadis, Gropius, Haar, Wagner). The fourth stage was one of *"Embracing of the Humanities"* (Friedrich, Gottmann, Higbee, Maass, Mead, Rudolph), the reaching out in many theoretical instead of practical directions. The fifth stage of late maturity brings more inner peace, with wide looks forward and backward; it is one of *"Philosophical Affinity and Receptivity"* (Buttimer, Hartke, Guelke, Toynbee, Tulippe, Ward). I believe I have slowly changed from a Promethean to a Protean and then to a Sisyphean geographer, and that my Apollonian base has been increasingly the foundation for a plethora of Dionysiac structural systems. From being a "monist" I have changed to being a "pluralist." I have been impacted by concepts all my life and have become increasingly aware of it. Shirley MacLaine said in *Out on a Limb*:

> I guess all people establish a set of unstated but understood rules of communication when they're together one on one. It's something you don't think about, but it's there and in operation until one of the two breaks what's established and attempts to go on to another level. (p. 237)

Most of the strongest humanistic injections I have received have been on this "one on one" basis, with mentors and friends, who reached me and lead me to another level of "verstehen." Thus I have become increasingly open to take that ride when propitious, or to initiate the impetus towards another level. Yet I also agree with Ralph Waldo Emerson that *true* fortitude of understanding consists in "*not* letting what we know to be embarrassed by what we *don't* know" (*The Skeptic*).

I wonder what the *sixth* stage now holds in store! Certainly it will be involvement in Future Studies, not from a technical or mechanistic vantage point, but from the humanistic approach of alternative future environments á la *Zukunftsforschung* now taught at some German universities. It is a logical extension of the application of geographical knowledge, involving such catchy topics as Societal Indicators (SI), Quality of Life (QOL), Technology Assessment (TA), Limits to

Growth (LG) and so on. Geographers, whether physical or cultural, young or old, urban or rural, experienced or untried, Malthusian or Cornucopian, must continue to open options, as technological fixes no longer suffice in this era of accelerating vulnerabilities. As Auguste Comte observed more than a century ago: "Savoir pour prévoir et prévoir pour pouvoir!" That is: "Know in order to foresee and foresee in order to be capable of doing!" During my circuitous journey, particularly during the last half century, I have learned that boundaries are less meaningful as the disintegration of national governance becomes more widespread. Racial tensions, terrorism, small wars between polarized factions and even wars "by proxy" are expanding. Problems virtually unknown at the beginning of my journey have now combined in recent years to create a *qualitatively* distinct class of *unavoidable* planetary problems; they are *not* only environmental, but also demographic, economic and political. Their solution is beyond the ability and the responsibility of individual nation-states. These "wicked" problems cannot be wished away and they are, above all, indifferent to military force! Meticulous and creative geographic thinking is needed more than ever. Geographic mind-sets effective at the time of World War II no longer suffice. The humanities have assisted and must continue to keep our discipline relevant and human(e), and will enlarge its scope and enhance its usefulness in the *management of change*. Who is better equipped to advise on speed limits of modernization to avoid dangerous collisions between the cultural resistance of nationalists, the religious opposition of reactionaries, the impoverished many and the enriched few? Accidents at dangerous intersections may be prevented by the application of sound geographic foresight, humanistically based and exhibited with some mediagenic panache.

One might ask whether there has not been one constant specific conceptual focus in this circuitous journey over the decades. Probably it is the constantly growing realization that in order to be "original" we must first explore *origins*. The examination of roots is an intensely humanistic affair. Boulez, Braudel, Darwin, Doxiadis, Freud, Fuller, Gaudi, Geddes, Giedion, Goethe, Gottmann, Gropius, Jung, Mumford, Ritter, Toynbee, Weber - here are some cases where original thinkers saw fit to explore origins. And this permeates creative thought in most directions. For instance, the ubiquitous topic of "boundaries" has always fascinated me, and has surfaced repeatedly in this account. It is symbolic of my journey. I suppose it started with my frightening crossing of international boundaries in my youth. It was stimulated by McBride in his talks on the Ecuador/Peru boundary disputes. It was fired up by Broek's lectures on Ratzel's concepts of *Grenzen*. It was extended by Hartshorne's discussions of functional thrust in his Political Geography course. It was stimulated by listening to Whittlesey and Stephen Jones debating the historical impact and the extent of compages. It achieved consistency via Doxiadis' ekistic logarithmic scale. It was enlarged by Gottmann's learned lectures on boundary decision-making, with special reference to the politics of circulation. It was impacted by Haushofer with his cynical views of boundaries as a national geopolitical tool. As indicated, many role models, exemplars and mentors have contributed to my strange love/hate affair with the concept. Yet the salient clue for my disaffection for boundaries can be found in the entry I provided for the "Thoughts on my Life" section of *Who's Who in the World* more than a decade ago:

> I have never worried about boundaries, whether geographical, intellectual, disciplinary, or any other type. One has to follow those avenues where one's intellectual curiosity points the way, even if these paths lead to entirely different territories. The world of reflective thinkers is inhabited primarily by splitters and drillers, but the lumpers and spreaders are increasing rapidly in this era of knowledge explosion, and I am a standard bearer of this salient force as I help to create better futures. (cf. recent update: *Who's Who in America*, Marquis Who's Who, Chicago, 43rd edition, 1984-1985, pp. 2383-84)

This is the reason for my constantly increasing affinity for the humanities, which ceaselessly intensifies on my journey. Humanistic thought has expanded the past boundaries of our discipline and will continue to dilute them in the future in the dimensions of space and time.

Acknowledgement

My warm thanks go to Anne Buttimer, University of Lund, who not only inspired me to put my thoughts on paper after our television interview in Geneva in August 1984 on *The Molding of Geographic Awareness* (DIA:G42), but who also gave this manuscript a critical reading. Unfortunately, her wise suggestion to start with the conclusions and to stress more the creative tension between Apollonian and Dionysian encounters, could not be realized because the contributions had already been formatted. However, I do wish to mention Buttimer's antidote to my expressed fear that the essay was too personal. She said: "Read Torston (Hägerstrand)'s parting shot in his essay in (my) *The Practice of Geography*, p. 256 . . . it could disarm some Pharisees!" Hägerstrand, describing his conversations with Buttimer, (*ibid*. Longman, 1983, pp. 255-56) states that "we agreed to try to initiate a dialogue on this urgent theme of a bridge between subjective experience and objective knowledge. We have tried several approaches. Autobiography is one which I have welcomed since its major theme, individual and context, fits so well into my general picture of the world. It is strange that some outside pressure was needed for me to see that the obvious point to begin when trying to link the outer and inner worlds is where you can look in both directions: *with oneself*." I agree with this basis for understanding in humanistic knowledge, geography included.

APPENDIX B

PETER H. NASH PUBLICATIONS

Peter H. Nash
School of Architecture
Department of Geography
and
School of Urban and Regional Planning
Faculty of Environmental Studies
University of Waterloo
Waterloo, Ontario

ALL publications listed and numbered in *chronological* order, and are identified by *types*, using the letters (A through 0) below.

List does NOT include some popular writings, commissioned entries for Chambers's Enclyclopaedia (59 articles), and Colliers Encyclopaedia (87 articles), unpublished professional papers, newspaper articles, and writings in the philosophy of science.

A Book

B Chapter in Book

C Editor of Book

D Major Article in Refereed Professional Journal

E Major Article in Non-Refereed Professional Journal

F Major Article in Book of Proceedings of Professional Society

G Minor Article in Refereed Professional Journal

H Minor Article in Non-Refereed Professional Journal

I Book Review in Refereed Professional Journal

J Condensation of Professional Paper Presented at a National Conference

K Condensation of Professional Paper Presented at an International Conference

L Thesis (on Microfilm)

M Dissertation (on Microfilm)

N Significant Professional Article in Major Newspaper

O Published Professional Report of Broad Significance

* Asterisk indicates listing in the *Catalogue of the Library of the Graduate School of Design, Harvard University*, G.K. Hall & Co., Boston, 1968.

1	1946	D	"Le Massif du Vercors en 1945: Etude sur les dévastations causées par l'Armée allemande dans une région alpine de la France et de leurs effets sur les traits géographiques." *Revue de Géographie Alpine*. Vol. XXXIV. February, 1946. pp. 87-101.
2	1947	L	*The Vercors Massif: A Geographical Analysis of a French Alpine Region*. University of California at Los Angeles Archives. Los Angeles, California. June, 1946. 223 pp.
3	1947	G	"The Swiss Emergency Cultivation Project, 1940-1945." *Geographical Review*. Vol. 37. No. 2. 1947. pp. 322-324.
4	1948	G *	"Montreal: An Outline in Urban Geography." *Geographical Review*. Vol. 38. No. 3. July, 1948. pp. 494-496.
5	1948	O *	*A Comprehensive Survey of Salient Problems in Urban Redevelopment Legislation*. Harvard Law School Archives. 1948. 192 pp. (VF NAC 1614 N)
6	1949	G *	"Design for Acoustics." *Journal of the American Institute of Planners*. Vol. XV. No. 3. Fall, 1949. p. 49.
7	1950	F *	"Techniques for Calculating Demographic Changes and Density Standards." *Planning: 1950*. Proceedings of the Annual National Planning Conference, Los Angeles, California. August, 1950. American Society of Planning Officials. pp. 68-79, with tables, chart. (NAC 42 Am 1950)
8	1950	I	"The Urban Pattern: City Planning and Design." *Journal of the American Institute of Planners*. Vol. XVI. No. 4. Fall, 1950. pp. 190-192.

9	1950	O	*General Plan for Boston: Preliminary Report*. December, 1950. Boston City Planning Board. 120 pp. (Co-author as Principal Planning Assistant.)
10	1951	O *	*Report to the Wakefield Town Planning Board*. Wakefield, Massachusetts. Published by the Town of Wakefield, Mass. 1951. 85 pp. (Detailed Recommendations for Pilot Plan.)
11	1953	G *	"Aspects of the Textbook Approach." *Journal of the American Institute of Planners*. Vol. XIX. No. 2. Spring, 1953. pp. 109-112.
12	1954	D *	"Planning as a Staff Function in Urban Management." *Journal of the American Institute of Planners*. Vol. XX. No. 3. Summer, 1954. pp. 136-147. (With James F. Shurtleff, Co-author.) (VF NAC 1520 S)
13	1955	H *	"Cuantos Mapas Necesita Una Ciudad?" (How Many Maps Does a Community Need?) *Servicios Publicos*. Vol. 2. No. 2. March, 1955. pp. 10-12, 26-29. (NAC 834 N)
14	1955	G	"Approaches to Community Analysis." *Journal of the American Institute of Planners*. Vol. XXI. No. 2-3. Spring-Summer, 1955. pp. 101-104.
15	1955	E *	"Como Formular Un Programa De Prioridad Para Mejoramiento de Calles." (How to Formulate a Priority Program for Street Improvements.) *Servicios Publicos*. Vol. 2. No. 5. September/October, 1955. pp. 14-15, 35-36.

16	1955	G	"American Geography: Inventory and Prospect." *Journal of the American Institute of Planners*. Vol. XXI. No. 4. Fall, 1955. pp. 170-173.
17	1955	C *	*Urban Renewal Report #1.* City of Medford, Massachusetts. 1955. 105 pp. Published by the City of Medford, Mass. P.H. Nash, Editor and Chief Contributor.
18	1955	N	*Medford in 1980*. 75th Anniversary Edition of the *Medford Mercury*, Medford, Massachusetts. June 1955. Sec. I. pp. 1-56.
19	1956	J	"Evolving Acceptance of Geographic Thinking in City Management, Community Planning, and Urban Renewal." *Annals of the Association of American Geographers*. Vol. XLVI. No. 2. June, 1956. pp. 266-267. (Abstract of Paper presented at Annual Meeting, Montreal, Quebec, April 1956.)
20	1956	D *	"The Case of the Case Study in City Planning and Municipal Management." *Journal of the American Institute of Planners*. Vol. XXII. No. 3. Summer, 1956. pp. 153-172. (With James F. Shurtleff, Co-author.) (NAC VF 1526 N)
21	1956	G *	"Man's Role in Changing the Face of the Earth." *Journal of the American Institute of Planners*. Vol. XXII. No. 4. Fall, 1956. pp. 256-259.
22	1957	O	*Confidential Reports on Urban Renewal Organization and Administration in Norfolk, Va., Little Rock, Ark., and Nashville, Tenn*. (With John A. Parker as Co-author.) Prepared for the Bureau of Public Administration, University of California. 89 pp. (Published in

			Condensed version under the sponsorship of the U.S. HHFA.)
23	1957	I	"Urban Land Use Planning." *The Professional Geographer.* Vol. IX. No. 6. November, 1957. pp. 38-39. (A Review of Volume by F. Stuart Chapin, Jr.)
24	1958	G *	"The Industrial Structure of American Cities." *Journal of the American Institute of Planners.* Vol. XXIV. No. 1. Spring, 1958. p. 47.
25	1958	G *	"Education for Planning: City, State, and Regional." *Social Forces.* Published by the Institute for Research in Social Science, University of North Carolina. Vol. 36. No. 4. May, 1958. pp. 383-384.
26	1958	O	*Methods and Techniques of Coupling Research and Teaching in the Preparation of Case Materials: The Case of the "Case Cadets" of the "Role of the Planner" Project at the University of North Carolina.* 48 pp. Planning Educators Meeting, National Academy of Sciences, Washington, D.C. May 18, 1958.
27	1958	M *	*The Responsibilities and Limitations of the Planning Director in a Council-Manager Form of City Government: An Exploratory Analysis Based on Case Studies of "Plan E" Cities in Massachusetts.* April, 1958. 999 pp. 2 vols. Harvard University Archives, Cambridge, Massachusetts. Microfilmed. (HCL HU 90. 7420 B)
28	1958	J	"Inventory or Prospect: An Analysis of the Responsibilities and Limitations of Geographic Field Work." *Annals of the Association of American Geographers.* Vol. 48. No. 3. September, 1958. p. 283. (Abstract of Paper presented at Annual

			Meeting, Los Angeles, California, August, 1958).
29	1959	J	"Interdisciplinary Approaches to the Study of Emerging Polynuclear Metropolitan Regions." *Annals of the Association of American Geographers*. Vol. 49. No. 2. June, 1959. p. 204. (Abstract of Paper presented at Annual Meeting, Pittsburgh, Pennsylvania, April, 1959.)
30	1959	D	"The Role of Value Judgements and Programming in Geography." *The Professional Geographer*. Vol. XI. No. 5. September, 1959. pp. 19-22.
31	1960	K	"Recent Trends in Applied Geography in the U.S.A.: Major Pulls and Selected Issues." *19th International Geographical Congress Abstract of Papers, Stockholm, Sweden*. 1960. pp. 210-212.
32	1960	J	*Dilemmas of Conflicting Objectives: For Whom Do We Plan?* American Institute of Planners Abstracts. 1960. 8 pp.
33	1960	A	*A Strategy for Improving Housing in Greater Cincinnati*. ACTION, Inc., New York City. 1960. 389 pp., with 26 maps. (Co-author.)
34	1961	O	*The Role of the Geographer in the Urban Renewal Process*. The Brookings Institute. Washington 6, D.C. April, 1961. 12 pp.
35	1961	F *	"People, Positions, and Pay." *Planning: 1961*. pp. 202-210. Proceedings of the Annual National Planning Conference, Denver, Colorado, May, 1961. American Society of Planning Officials. (Topic: *Planning Administration: Past, Present, and Future*.)

36	1961	D *	"For Whom Do We Plan?" *Ekistics: Reviews on the Problems and Science of Human Settlements.* Vol. 13. No. 75. January, 1962. pp. 53-56.
37	1962	D *	"A Geographer Suggests How His Profession Can Help in Renewal Job." *Journal of Housing.* Vol. 19. No. 2. February, 1962. pp. 83-86. (Fourth of a series of five articles written for the "Committee on Problems of the American Community" of The Brookings Institution, Washington, D.C.)
38	1962	D *	"Legislative Reapportionment, Urban Planning, and the Supreme Court." *Journal of the American Institute of Planners.* Vol. XXVIII. No. 3. August, 1962. pp. 145-151, with tables. (With Richard L. Strecker as Co-author.)
39	1962	J *	"A Set of Maxims for Planners: Dilemmas of Conflicting Objectives." *Proceedings of the 43rd Annual Conference: American Institute of Planners.* October, 1960. (Published 1962.) pp. 126-133.
40	1962	J	"Action System Characteristics in Area Analysis." *Annals of the Association of American Geographers.* Vol. 52. No. 3. September, 1962. pp. 351-352. (Abstract of Paper presented at the 57th Annual Meeting in Miami Beach, Florida.)
41	1962	K	"La géographie et l'aménagement regional." *Colloque National de Géographie Appliquée.* Centre National de la Recherche Scientifique. Paris. 1962. pp. 180-193, 199.
42	1963	G	"The Challenge of Ekistics: Duties Ahead." *Annals of the Association of American Geographers.* Vol. 53. December, 1963. p. 613.

43	1963	J	"Megalopolis and Megageography." *Annals of the Association of American Geographers*. Vol. 53. December, 1963. pp. 612-613. (Abstract of paper at "Megalopolis Symposium" at Denver, Colorado; September, 1963.)
44	1964	K	"Derivation and Utilization of Social Value Parameters for Decentralized Cost-Benefit Analysis of Alternative Social Policies: Economic Implications of Index of Desirability Via Objective Functions." *Proceedings of the 27th World Congress, I.F.H.P., Jerusalem, Israel.* (International Federation for Housing and Planning.) 1964. pp. 213-235.
45	1964	D *	"A Task-Force Approach to Replace the Planning Board." *Journal of the American Institute of Planners*. Vol. 30. No. 1. pp. 10-26. February, 1964. (With Dennis Durden as Co-author.)
46	1964	K	"Applications of Megageographical Concepts." *Abstracts of Papers: 20th International Geographical Congress, London, England.* Nelson & Sons Ltd. 1964. pp. 170-171.
47	1965	I *	"The Federal Bulldozer." *Boston College Industrial and Commercial Law Review*. Vol. VI. No. 4. Summer, 1965. pp. 972-976.
48	1965	J	"The Responsibilities and Limitations of the Graduate Dean in the Dilution of Disciplinary Boundaries." *Proceedings of the Fifth Annual Meeting of the Council of Graduate Schools in the United States.* Washington, D.C. December, 1965. pp. 66-71.

49	1965	D *	"Public Planning Boards: Abolition or Systematic Proliferation?" *Boston College Industrial and Commercial Law Review*. Vol. VII. No. 1. Fall, 1965. pp. 37-61. (VF NAC 1535 N)
50	1966	E	"Building Bridges Among Graduate Schools." *Proceedings of the New England Conference on Graduate Education*. 23rd Annual Meeting. May, 1966. pp. 34-45. Clark University, Worcester, Massachusetts.
51	1966	B	"Quelques idées sur l'origine et l'évolution de la géographie appliquée aux Etats Unis." *La Géographic Appliquée Dans Le Monde*. Actes de la Réunion à Prague, September, 1965. Miroslav Strida, Editor. Academia, Nakladatelstvi Ceskoslovenske Akademie Ved 1966. pp. 30-36. (Some Thoughts on the Origins and Development of Applied Geography in the United States.)
52	1966	B	"The Interdisciplinary Seminar for Professional Geographers." *Applied Geography in the World*. Proceedings of the Prague Meeting, September, 1965. Miroslav Strida, Editor. Academia, Nakladatelstvi Ceskoslovenske Akademie Ved 1966. pp. 129-136. (Seminaire interdisciplinaire pour les geographes professionels.) Also, in same issue: "Problems of Geographic Applications in Countries with Free Enterprise Economies."
53	1967	B *	"Introduction" (pp. iv-vi) and "Opening Remarks" (p. 1. *et seq.*) *Proceedings of the Second International Meeting, Commission on Applied Geography, International Geographical Union*. August, 1966. A.A. Michel, Editor. University of Rhode Island, Kingston, Rhode Island. (NAC 7170 In7)

54	1967	B *	"From 'Heimatkunde' to 'Ekistic Grid': Attitudinal Polarities." (Address presented in honour of President Omer Tulippe on his Retirement.) *Proceedings of the Second International Meeting, Commission on Applied Geography, International Geographical Union.* August, 1966. A.A. Michel, Editor. University of Rhode Island, Kingston, Rhode Island. pp. 203-205. (NAC 7170 In7)
55	1967	K *	"Note on the Development of a General Plan for Education." *Ekistics: Reviews on the Problems and Science of Human Settlements.* Vol. 24. No. 143. October, 1967. pp. 361-363, with port. (Also: Need for Research and Experimentation.)
56	1967	B *	"Pressures Brought by Urban Renewal" in *Metropolis on the Move: Geographers Look at Urban Sprawl.* (Edited by Gottmann, Jean and Harper, Robert A.) John Wiley & Sons, Inc., New York. 1967. pp. 43-57.
57	1967	D *	"Multiple Therapies Plus Programed Kinetogenesis: A Shading Difference." *Ekistics: Reviews on the Problems and Science of Human Settlements.* Vol. 24. No. 145. December, 1967. pp. 454-456, with port. Full text also in: *Proceedings of the International Seminar on Ekistics.* Document B. No. 14, pp. 1-9. Athens Center for Ekistics. 1967.
58	1967	K	"The Human Community." *Ekistics: Reviews on the Problems and Science of Human Settlements.* Vol. 24. No. 145. December, 1967. Summary of Research Discussion. pp. 508-512.
59	1967	E	"How Can We Make Better Teachers Out of Our Ph.D.'s?" and "Report of Secretary-Treasurer." *Proceedings of the New*

			England Conference on Graduate Education. 24th Annual Meeting. Brandeis University, Waltham, Massachusetts. May, 1967, pp. 28-34 and 49-50.
60	1968	B *	"Applied Geography and Ekistic Expertise." *Colloque International de Géographie Appliquée: Comptes Rendus*. Universite de Liège, 1967. Vol. 48. Liège, Belgium. 1968. pp. 53-62. (VF NAC 1321 N)
61	1968	B	"The Three Major Problems of Training Applied Geographers." *Colloque International de Géographie Appliquée: Comptes Rendus*. Universite de Liège, 1967. Vol. 48. Liège, Belgium. 1968. pp. 339-343.
62	1968	G *	"Planning, Geography, and Ekistics." *Newsletter: Bureau of Government Research*, University of Rhode Island, Kingston, Rhode Island. Vol. IX. No. 5. January, 1968. pp. 1-4. (VF NAC 1321 N)
63	1968	B *	"Two Contrasting Systems of Geographic Applications." *Mélanges de Géographie: Physique, Humaine, Economique, Appliquée*. Editions J. Duculot, S.A., Gembloux, 1968. Vol. II. pp. 518-522.
64	1968	K	"Music Regions and Geography." *Abstracts of Papers Presented at the Twenty-first International Geographical Congress*, New Delhi, India. December, 1968. Abstract No. 770. pp. 316-317.
65	1968	B	"The Client of the Applied Geographer." *Proceedings of International Land Use Seminar, Aligarh Muslim University*, Aligarh, India. November, 1968. pp. 48-72.

66	1968	D	"Music Regions and Regional Music." *The Deccan Geographer*. (Semi-annual Journal of the Deccan Geographical Society.) Secunderabad, A.P., India. Vol. VI. No. 2. July-December, 1968. pp. 1-24.
67	1969	I	"The Regional City." *Economic Geography*. Vol. 45. No. 2. April, 1969. pp. 180-181.
68	1969	D	"Ekistics Theory and Practice: The Programmed Development of Settlement Patterns." *Rhode Island Business Quarterly*. Research Centre in Business and Economics, College of Business Administration, University of Rhode Island. Vol. 5. No. 3. September, 1969. pp. 13-20.
69	1969	O	*The General Environment for Contract Research and Priority Areas*. (Including Theoretical Implications of the Battelle Cleveland Experience and the New Communities Act of 1968.) Battelle Memorial Institute. November, 1969. 43 pp.
70	1970	O	*Report to Committee on Military Geography*. Advisory Board on Military Personnel Supplies of the Division of Engineering. National Research Council. National Academy of Sciences. Washington, D.C. No. LVIII. 1970. 110 pp.
71	1970	D	"Urban Public Policy Participation Networks." *Urban and Social Change Review*. Institute of Human Sciences, Boston College, Chestnut Hill, Massachusetts. Vol. III. No. 2. Spring, 1970. pp. 15-20. (With Howard H. Foster, Jr. and D. Barlow Burke, Jr., Co-authors.)

72	1970	K	"The Relationship of Ekistic Units, Movement Networks, and Communications to Other Life Style Phenomena." *Delos Symposion* (Delos Eight). Document B. No. 18. Athens Center for Ekistics. July, 1970. 3 pp.
73	1970	K	"The Owl and the Pussycat: Network of Citizen Participation in the Development of Public Policy at Various Ekistic Levels." *Delos Symposion* (Delos Eight). Document B. No. 41. Athens Center for Ekistics. July, 1970. 3 pp.
74	1970	K	"Life Styles, Owls, Pussycats, Landscapes and the Ekistic Logarithmic Scale." *Ekistics: Reviews on the Problems and Science of Human Settlements*. Vol. 30. No. 179. October, 1970. p. 342, *et. seq*.
75	1971	K	"II IV Simposio Della Commissione Di Geografia Applicata Dell'Unione Geografica Internazionale." *Bollettino Della Societa Geografica Italiana*. Rome 1971. pp. 489-508. (With Carlo Della Valle.)
76	1971	B	"Futurism: The Newest Stage in Geographical Imagination." *Geography and Long Term Prospects*. (C.G.A. #4). Michel Phlipponneau, Editor. Presses Universitaires de Bretagne, Rennes, France, 1971. Editions Coconnier. Chapter I. pp. 37-43.
77	1971	B	"Environmental Umbrella for Geographic Pedagogy." *Geography and Long Term Prospects*. (C.G.A. #4). Michel Phlipponneau, Editor. Presses Universitaires de Bretagne, Rennes, France. 1971. Editions Coconnier. Chapter III. pp. 169-177.

78	1971	B	"Planning, Geography, and Ekistics." *Studies in Applied and Regional Geography*. Edited by Mohammad Shafi and Mehdi Raza. 1971. Sangam Press, New Delhi, India. pp. 45-56.
79	1972	O	"Environmental Iconophobia via Unrealistic Utopianism: A Critique of Implied General Goals of Applied Planning Methodology." *Proceedings Paper #4*. American Society of Planning Officials. 38th National Planning Conference, Detroit, Michigan. April, 1972. 12 pp.
80	1972	H	"Planning Education." *The Bulletin of the Association of Collegiate Schools of Planning*. Vol. 10. No. 2. Summer, 1972. pp. 8-10. (Implications of American Society of Planning Officials *Proceedings Paper #4*.)
81	1972	J	"Iconophobia and Environment." Proceedings of the New England - St. Lawrence Valley Geographical *Society*. Durham, New Hampshire. October, 1972. pp. 84-85. (Abstract).
82	1972	O	"Planning Icons: Apollonian and Dionysian." *Confer-In Paper #31*. American Institute of Planners, 55th Annual Conference, Boston, Massachusetts. October, 1972. 15 pp.
83	1972	K	"Correlates of Environmental Image Dogma: Kant, Nietzsche, Ratzel, Gropius." *Wissenschaftliche Mitteilungen der Goethe Universität*. Frankfurt am Main. No. 286. December, 1972. pp. 4-11. (With a summary in German on pp. 10-11).
84	1972	D	"Unitariness and Vulnerability." *Geographical Outlook*. Vol. VIII. 1971-72. Department of Geography, Ranchi University, Ranchi, India. pp. i-x.

85	1973	D	"Iconophobia and Geography." *The Geographer*. The Aligarh Muslim University Geographical Society. Aligarh - 202001, India. Vol. XX. No. 1. January, 1973. pp. 1-15.
86	1973	D	"Antidotes to Academiasis and Diplomania: Recent Trends in Geography Education." *Ekistics: Reviews on the Problems and Science of Human Settlements*. Vol. 35. No. 210. May, 1973. pp. 278-281.
87	1973	J	"Geography, Public Policy and Inconophobia." Abstracts of Submitted Papers - Resumes. Canadian Association of Geographers. Lakehead University, Thunder Bay, Ontario. May 27-31, 1973. pp. 45-46.
88	1973	G	"Icons, Iconography and Iconophobia." *The Bulletin of the Association of Collegiate Schools of Planning*. Vol. XI. Number 2. Summer, 1973. pp. 6-10.
89	1973	D	"The Computer in Applied Geography." *The Geographer*. The Aligarh Muslim University Geographical Society. Aligarh - 202001, India. Vol. XX. No. 2. July, 1973. pp. 87-97.
90	1973	D	"A Plea for the Rejection of 'Rejection'." *The Professional Geographer*. Vol. XXV. No. 3. August, 1973. pp. 211-213.
91	1973	D	"Die Universität von Waterloo - Eine Neuartige Ausbildungsstätte." *Umschau in Wissenschaft und Technik*. Vol. 73. No. 18. September, 1973. pp. 559-562. (With Peter C. Brother, Co-author. Includes photos and chart.)

92	1973	I	"Arts of the Environment." *Planning: The ASPO Magazine*. Vol. 39. No. 9. October, 1973. pp. 35-36. (Review of Gyorgy Kepes volume).
93	1973	B	"Applications of Geomorphological Field Work in the Canadian High Arctic." *Applied Geography and the Human Environment*. (Proceedings of the Fifth International Meeting, Commission on Applied Geography, International Geographical Union, August 1972.) Richard E. Preston, Editor. (Geography Publication Series No. 2). University of Waterloo, Ontario, Canada. 1973. pp. 45-52.
94	1973	B	"The Use of the Computer in Environmental Studies." *Applied Geography and the Human Environment*. (Proceedings of the Fifth International Meeting, Commission on Applied Geography, International Geographical Union, August, 1972.) Richard E. Preston, Editor. (Geography Publication Series No. 2.) University of Waterloo, Ontario, Canada. 1973. pp. 98-105. (With Terence McL. Semple, Co-author.)
95	1973	B	"Emerging Trends in Environmental Training of Geographers in North America: Antidote to Academiasis and Diplomania." *Applied Geography and the Human Environment*. (Proceedings of the Fifth International Meeting, Commission on Applied Geography, International Geographical Union, August, 1972). Richard E. Preston, Editor. (Geography Publication Series No. 2.) University of Waterloo, Ontario, Canada, 1973. pp. 318-329.

96	1973	B	"Unitariness and Vulnerability: Reflections of a Non-Participant on the Implications of the United Nations 'Conference on the Human Environment' held in Stockholm, June 5-16, 1972." *Applied Geography and the Human Environment.* (Proceedings of the Fifth International Meeting, Commission on Applied Geography, International Geographical Union, August, 1972.) Richard E. Preston, Editor. (Geography Publication Series No. 2.) University of Waterloo, Ontario, Canada. 1973. pp. 347-355.
97	1973	D	"Iconophobia and Environment." *Geographical Outlook.* Vol. IX. 1972-73. Department of Geography, Ranchi University, Ranchi, India. pp. 1-13. (Expansion of Bibliography Item #81.)
98	1974	J	"Polyphony, Monogenesis, Topophilia, Iconophobia, Troubadours, Organs, Rock, and Sex." *Abstracts of Annual Meeting, The Canadian Association of Geographers: 1974.* University of Toronto/York University. 1974. pp. 20, 142-143.
99	1974	B	"Futurism and Geographical Imagination." *Man and Environment.* Studies in Geography in Hungary, 11. Geographical Research Institute. Hungarian Academy of Sciences. Academiai Kiado. Budapest. 1974. pp. 35-39.
100	1974	H	"Arts of the Environment." *Contact: Bulletin of Urban and Environmental Affairs.* Vol. 6. No. 3. June, 1974. pp. 23-24. University of Waterloo, Ontario.
101	1974	K	"Educating for Enlightened Participation in Settlement Decisions: A Needed Action (With Special Reference to Categories and Themes in Environmental Post-Secondary

			Education)," *Delos Symposion* (Delos Eleven). Document B. No. 18. Athens Center for Ekistics, July 1974. 4 pp.
102	1974	K	"Harmonic Topophilia and Apollonion," *Delos Symposion* (Delos Eleven). Document B. No. 25. Athens Center for Ekistics. July, 1974. 2 pp.
103	1974	I	"Compact City: A Plan for a Liveable Urban Environment." *Canadian Geographical Journal*. Vol. 89. No. 1 & 2. July/August, 1974. p. 53. (A Review of Volume by George B. Dantzig and Thomas L. Saaty.)
104	1974	H	"Categories of Environmental Education." *Contact: Bulletin of Urban and Environmental Affairs*. Vol. 6. No. 6. December, 1974. Occasional Paper No. 15. ("International Perspectives on the Education of Architects and Planners.") pp. 1-3.
105	1974	K	"Music and Cultural Geography." *Abstracts of New Zealand Regional Conference of the International Geographical Union*. Massey University, Palmerston North, New Zealand. December 4-11, 1974. (Sponsored by The Royal Society of New Zealand.)
106	1974	I	"Topophilia: A Study of Environmental Perception, Attitudes and Values." *Canadian Geographical Journal*. Vol. 89. No. 6. December, 1974. p. 46. (A Review of Volume by Yi-Fu Tuan.)
107	1975	I	"Another Bottle of Perfume!" (A Review of "Compact City" by Dantzig and Saaty.) *Contact: Bulletin of Urban Environmental Affairs*. Vol. 7. No's. 3/4. June/July, 1975. Occasional Paper No. 18. pp. 42-43.

108	1975	D	"Music and Environment: An Investigation of Some of the Spatial Aspects of Production, Diffusion, and Consumption of Music." *Journal of the Canadian Association of University Schools of Music* (JCAUSM/JACEUM). Vol. V. No. 1. Spring, 1975. pp. 42-71, plus 16 unpaged maps. (Includes "Appendix": Environmental Reference Units for Musical Activities.)
109	1975	B	"Iconophobia and Environment." *Les Régions Qu'il Faudrait Faire* (Building Regions of the Future.) Notes et Documents de Recherche. No. 6, December, 1975. Départment de Géographie, Université Laval, Québec. (167 pp.) Bélanger, Marcel and Janelle, Donald G., Editors. pp. 33-45. (Revision of Bibliography Item #81 for Second Conference "The Geography of the Future," University of Montreal.)
110	1975	D	"Music and Cultural Geography." *The Geographer*. The Aligarh Muslim University Geograpical Society. Aligarh - 202001, India. Vol. XXII. No. 1. January, 1975. pp. 1-14.
111	1975	B	"Iconophobia and Environment." *Applied Geography Papers*. Vol. 1. Joplin, Missouri. (Mary Megee, Editor.) 1975. pp. B-1 to B-9. (Revision of Bibliography Item #97.)
112	1975	D	"Isomedians and Human 'Watersheds'." *Geographical Outlook*. Vol. XI. 1974-75. Department of Geography, Ranchi University, Ranchi, India. pp. i-x. (Expansion, with graphs, of brief summary published in *Abstracts of 61st Annual Meeting, Association of American Geographers: 1965*. The Ohio State University, Columbus, Ohio. p. 85.)

113	1975	F	"Perceptions of Megalopolis." *Great Lakes Megalopolis: From Civilization to Ecumenization*. (Volume based on the Proceedings of the Great Lakes Megalopolis Symposion held in Toronto City Hall, March 24-27, 1975, Canada. 118 pp.) Ministry of State, Urban Affairs, Canada. (Alexander B. Leman and Ingrid A. Leman, Editors.) pp. 10-13, *et. seq*.
114	1976	J	"Mosaic, Incubator, Hinge, and Icon: Birth Pangs of a Higher Urban System." *Abstract of Annual Meeting, The Canadian Association of Geographers: 1976*. Université Laval. May, 1976. pp. 93-95.
115	1976	D	"Injecting the 'Future' into Environmental Studies." *Transition: Quartery Journal of the Socially and Ecologically Responsible Geographers*. Vol. 6. No. 3. Fall, 1976. pp. 9-12. (Summary of 7 pp. multilithed paper presented at the Founding Conference on "Future Studies in Canada" of the Canadian Association of Future Studies, University of Western Ontario, London, February 7, 1976.)
116	1976	D	"Realistic Conceptual Issues Confronting the Great Lakes Megalopolis: The Inevitable, the Desirable, and the Feasible." *The East Lakes Geographer*. Vol. XI. June, 1976. pp. 45-58.
117	1976	G	"Injection of Topophilia Via Historical Geography: Reflections on Harvard's 'The Boston Region' Field Course As Taught by Derwent Whittlesey." *Proceedings of the New England - St. Lawrence Valley Geographical Society*. Association of American Geographers. Vol. VI. October, 1976. pp. 59-63.

118 1977 D "Environmental Quality" (Proceedings of the Fourth World Congress of Engineers and Architects) and "Territorial Settlements and Settlements in Territories." *ITCC Review.* (International Technical Cooperation Centre, Association of Engineers and Architects, Tel-Aviv, Israel.) Vol. VI. No. 2. (22), April, 1977. pp. 57-66.

119 1977 J "Alfred and Friedrich and Pierre and René: Shades of 'Heimat' and Degrees of 'Nationness'." *Programme and Abstracts of the Canadian Association of Geographers.* University of Regina, Saskatchewan. June, 1977. pp. 27-31.

120 1977 K "Territorial Settlements and Settlements in Territories: Reflections of a Participant on the Geographic Implications of the United Nations 'Conference on Human Settlements' (HABITAT) held in Vancouver, Canada, May 27 to June 12, 1976." *Contact: Journal of Urban and Environmental Affairs.* Vol. 9. No. 2. Summer, 1977. pp. 121-136. University of Waterloo, Ontario. (Summary of paper presented to the Commission on Applied Geography, International Geographical Union, Tbilisi, Georgian S.S.R., July 1976.)

121 1977 I "A Comparative Atlas of America's Great Cities: Twenty Metropolitan Regions." *Association of Canadian Map Libraries Bulletin.* (Association des Cartotheques Canadiennes.) No. 25. October, 1977. pp. 38-41. (Printed by Public Archives of Canada, Ottawa, Ontario.) Review of atlas of 503 pp. published by the Association of American Geographers and edited by Ronald Abler.

122	1977	D	"Territorial Settlements and Settlements in Territories." *GeoJournal: International Journal for Physical, Biological and Human Geosciences.* Akademische Verlagsgesellschaft. Wiesbaden, F.R. Germany. Vol 1. No. 4. 1977. pp. 7-14. (Applied Version of Bibliography Item #120 for Special Issue of GeoJournal for Working Group: Applied Aspects of Geography Annual Meeting #7, September, 1977.)
123	1977	D	"'Future Studies' and the Geo-Bio-Sciences." *GeoJournal: International Journal for Physical, Biological and Human Geosciences.* Akademische Verlagsgesellschaft, Wiesbaden, F.R. Germany. Vol. I. No. 6. 1977. pp. 13-16. (Adapted condensation of "Theme Paper" presented at the Founding Conference of the Canadian Association for Future Studies at the University of Western Ontario.)
124	1978	B	"Classical Music" in *The Sounds of People and Places: Readings in the Geography of Music.* Edited by George O. Carney, Oklahoma State University. University Press of America, Washington, D.C., 20023. pp. 1-52. 1978. (Condensed version of Bibliography Item #108, constituting Part One of a 336 pp. vol.)
125	1978	J	"From Promethean Faith to Protean Belief and Sisyphean Affirmation: Stages of 'Futures' Learning in Geographic Education and Research." *AAG Program Abstracts: New Orleans, 1978.* (Edited by Robert C. West and Clarissa Kimber.) April, 1978. pp. 104-105.

126 1978 J "Versatile Whittlesey: Influences On and From a Sisyphean Synthesizer." *CAG Volume of Abstracts* (*ACG Volume des Résumés*). The Canadian Association of Geographers. London. May, 1978. (Department of Geography, The University of Western Ontario.) p. 70. (Full text of 9 pp. distributed at London meeting.)

127 1978 D "On Futurizing Planning Education." *The Bulletin of the Association of Collegiate Schools of Planning*. Vol. XVI. Number 2. Summer, 1978. pp. 1-6, 16.

128 1979 I "Ecology and Ekistics." *Geographical Review*. Vol. 69. No. 1. January, 1979. pp. 109-110.

129 1979 J "Codign Essence and Influence: Normative and Strategic Tasks in Presenting 'The Nature of Geography'." *Program Abstracts: 75th Anniversary Meeting of the Association of American Geographers*. (John R. Mather, Editor.) April, 1979. p. 111.

130 1979 J "'Idiographic' and 'Nomothetic' Approach Patterns: Re-enforcing Feed-back Loops, Codign Staging Areas and Multiple Bridgeheads." *Abstracts: Annual Meeting '79* (*Résumé: Réunion Annuelle '79*). The Canadian Association of Geographers. Victoria. May, 1979. (Department of Geography, University of Victoria.) p. 31. (Full text of 10 pp. distributed at Victoria meeting.)

131 1979 I "The Sounds of People and Places." *Journal of Geography*. Vol. 78. No. 5. September/October, 1979. p. 206.

132 1979 H "Academic Versus/And Applied: Reply to the President." *Association of American Geographers Newsletter*. Vol. 14. No. 9. 1979. (Letter to the Editor.) Reprinted by

			permission of the Association of American Geographers in *Geography Evolving*, P.H. Nash and G. McBoyle. University of Waterloo. Third Edition, *et. seq.*
133	1980	G	"An Assortment of Assiduous Assessments and Assumptions for Assuagement and Assurance." *Journal of the American Planning Association*. Vol. 46. No. 2. April, 1980. pp. 213-215.
134	1980	K	"Geographical Aspects of Manufacturing and Colonization in Outer Space." *Symposium on the Applied Aspects of Geography*. International Geographical Union. Yokohama, Japan. August 25-29, 1980. Session I. 21 pp.
135	1980	I	"Urban Development in the U.S.A. and Hungary." *GeoJournal: International Journal for Physical, Biological, and Human Geosciences*. Akademische Verlagsgesellschaft. Wiesbaden, F.R. Germany. Vol. 4. No. 4. 1980. p. 388. (A Review of Volume by G. Enyedi, Editor, Académiai Kiadó.)
136	1981	J	"Values, Visions, and Vibrations: A Comparison of Geographic Epistemologies." *Program Abstracts of the Association of American Geographers: 1981.* (Kane, Phillip and Hornbeck, David, Co-Editors.) Los Angeles, Calif. April, 1981. p. 42. (Full text of 10 pp. plus chart distributed at Los Angeles meeting and available through Social Science Education Consortium, Inc., E.R.I. Center, Boulder, Colo.)
137	1981	D	"Versatile Whittlesey: Influences on and from a Sisyphean Synthesizer." *History of Geography Newsletter*. Association of American Geographers. Vol. 1. pp. 15-20. (Revision of Bibliography Item #126.)

138	1982	J	"Flows of Potential Power: Metagame Analysis of Issues in Radissonia, Quebec." *A.A.G. Program Abstracts*, San Antonio, Texas, April, 1982. p. 288. (With Christian M. Dufournaud; abstract of joint paper on James Bay Hydro-electric Project.)
139	1982	K	"The Philosophical Base and Source of Applied Geography in an Inter-Epistemological Spectrum." *Seminário sobre: O Estado E Os Desequilîbrios de Desenvolvimento Regional* (State of Regional Development). Grupo de Trabalho de Aspectos Aplicados da Geografia; Uniao Geográfica Internacional, Comissáo Nacional do Brasil. Recife, Brazil. 1982. (Texto de Comunicac<,oes Apresentadas; incl. Comparison Chart.)
140	1982	K	"Lacunate Leitmotifs or Laudable Lollipops?" *Simpósios E Mesas Redondas.* Latin American Regional Conference. I.G.U. Vol. II. 1982. Supplement. Sec<,ao Principal. Rio de Janeiro. I.B.G.E. (with Jane Schneider Pereyron Mocellin.) ISBN 85-240-0060-1.
141	1982	G	"Magnum Opus Via Sisyphean Confrontation: Moewes' Heroic Attempt to Bring Consistency to the Spatial Specialty." *GeoJournal: International Journal for Physical, Biological and Human Geosciences and their Application in Environmental Planning and Ecology.* Akademische Verlagsgesellschaft, Wiesbaden. Vol. 6. No. 5. pp. 493-494. (Critical Essay.)
142	1982	J	"The Non-Demise of Geography: Stays of Execution." *Proceedings Annual Meeting.* East Lakes Division of the Association of American Geographers. Western Michigan

			University, Kalamazoo. November, 1982. p. 5.
143	1982	I	"Basic Issues in the Design of Living Spaces: A New Comprehensive Approach from Germany." *Environments: A Journal of Interdisciplinary Studies*. University of Waterloo. Vol. 14. No. 3. pp. 60-62. (Adaptation of Bibliography Item #140.)
144	1982	I	"Themes in Geographic Thought." *Annals of the Association of American Geographers*. Vol. 73. No. 1. March, 1983. pp. 169-173.
145	1983	G	"A Two-Headed Phoenix Can Rise from the Ashes: Suggestions for the Redirection of Efforts of the I.G.U. Working Group on Applied Geography for the Post - 1984 Period." *GeoJournal: International Journal for Physical, Biological and Human Geosciences*. Akademische Verlagsgesellschaft. Wiesbaden. F.R. Germany. Vol. 7. No. 3. 1983. pp. 213-215.
146	1983	D	"Dr. Moewes Wrestles with the Future." *Transition: Quarterly Journal of the Socially and Ecologically Responsible Geographers*. Spring, 1983. Vol. 13. No. 1. pp. 20-23.
147	1984	I	"The Origins of Academic Geography in the United States." *Journal of Historical Geography*. January, 1984. Vol. 10. No. 1. pp. 93-95. (Review of volume by Brian W. Blouet, editor. Archon Books, 1981.)
148	1984	B	"Approaches and Methodology" (Section I) in P. Pandey (editor), *Modern Geographic Trends*. (Ranchi University, Ranchi, India). 1984. pp. 3-11.

149	1984	O	"From Philosophical Source to Policy Estuaries: Flows of Applied GeographERS." *IGU Discussion Paper*. 12 pp. mimeographed. (Mary Magee, editor.) Working Group of Applied Aspects of Geography of the International Geographical Union, National Policy Sessions. April 22, 1984. Washington, D.C.
150	1984	I	"The Coming of the Transactional City." *The Professional Geographer*. November, 1984. Vol. 36. No. 4. pp. 505-507. (Review of volume by Jean Gottmann, University of Maryland Institute for Urban Studies.)
151	1984	B	"Suggestions for the Redirection of Efforts of the I.G.U. Working Group on Applied Geography for the Post-1984 Period," *Aplikovaná Geographie: Sborník Prací 4*. Brno 1984. pp. 159-167, (and Resolution.) Edited by: Ceskoslovenská Akademie Ved Geografický Ústav. (Miroslav Strida & Jan Kára, Editors.)
152	1984	B	"Camelot Re-evaluated" in John Landis (editor), *The Future of National Urban Policy*. (College of Resource Development, The University of Rhode Island.) 1984. pp. 3-7.
153	1984	H	"Überlegungen zur Lieblingsumgebung und Lieblingslandschaft des Menschen." *Geographisches Kolloquium der Eberhard-Karls-Universität Tübingen*. Tübingen, West Germany (Condensation of Public Lecture at Geographical Institute), November, 1984. p. 5, *et. seq.*
154	1985	J	"Acral Geography Eschewed: A Prototype." *AAG Program Abstracts*, Association of American Geographers, Detroit. 1985. #280. (Geographic Thought II.)

155	1985	I	"The Spirit and Purpose of Planning." *Environments: A Journal of Interdisciplinary Studies*. Vol. 17. No. 1. May, 1985. pp. 75-77. (Review of Second Edition of Michael Bruton's "The Built Environment" Series.)
156	1985	J	"Raoul Blanchard: Promethean Progenitor and Peripatetic Professor." *Canadian Association of Geographers 85th Annual Meeting*. Université du Québec à Trois-Riviéres. May, 1985. pp. 65-66. (Abstract of 8 page mimeographed paper.)
157	1985	F	"Ekistics: One of the Ten 'Types' in an Epistemological Spectrum." *Proceedings of the International Seminar on Ekistics*. (Panayis Psomopoulos, Editor.) Document C. No. 1. 1985. pp. 6, plus chart. Athens Center of Ekistics, Athens, Greece.
158	1985	K	"Sarcophagus and Cenotaph: The Thanatology of Disciplinary Practices in Academia." *Proceedings of the International Seminar on Ekistics*. (Panayis Psomopoulos, Editor.) Document C. No. 2. 1985. pp. 8. Athens Center of Ekistics, Athens, Greece.
159	1985	F	"Systematic Synchronized Synergetic Synthesis as a Syndicated Symbol." *Proceedings of the International Seminar of Ekistics*. (Panayis Psomopoulos, Editor.) Document C. No. 3. 1985. pp. 4. Athens Center of Ekistics, Athens, Greece.
160	1986	D	"Favorite Landscapes." *Transition: Quarterly Journal of the Socially and Ecologically Responsible Geographers*. Vol. 15. No. 4. Winter, 1986. pp. 26-29.

161 1986 A *Dialogue Project* (Transcript Series G 42: The Molding of Geographic Awareness.) 1986. Reprocentralen, Lund, Sweden. 25 pp. (Also available on 3/4" U MATIC video-cassette and 1/2" VHS or Betamax via INTER-VIEWS, Lund University Library.) Recorded at University of Geneva, Switzerland; August, 1984.

162 1986 B "The Making of a Humanist Geographer: A Circuitous Journey" in *Geography and Humanistic Knowledge.* (Chapter 1.) Waterloo Lectures in Geography: Volume 2. (Edited by Leonard Guelke.) Department of Geography Publication Series No. 25. 1986. pp. 1-22.

163 1986 G "Favourite Landscapes." *Environments.* Vol. 18. No. 1 & 2. 1986. pp. 77-79. (Elaboration of Bibliography Item #160.)

164 1986 I Land-Use Planning: From Global to Local Challenge. (Review of volume by Julius Gy. Fabos.) *Town Planning Review.* Liverpool University Press. Vol. 57. No. 3. July, 1986. pp. 332-334.

165 1986 I "The Good Life." *The Canadian Geographer.* Vol. 30. No. 4. Winter, 1986. pp. 378-379. (Review of volume by Yi-Fu Tuan, University of Wisconsin Press.)

166 1986 F "La pera es deliciosa; las mariposas son bonitas. (Receptivity to external influences from Goethe's mid-life crisis to hedonistic casuistry)." *U.G.I., Facultat de Geografia i Història de la Universitat de Barcelona* (*U.B.*), 23 pp. 1986. (Expanded printed version of lecture given at University of Barcelona.)

167	1986	D	"Mark I Role Model and Inspiring Catalyst: Some Characteristics from a Perspective of Four Decades." *Cahiers de Géographie du Québec*, Vol. 30. No. 80. September, 1986. pp. 151-161.
168	1986	K	"A Spectrum of Mare Nostrum Concepts: Some Trenchant Labels from Immortals and Adopted Ideas from Mortals." *Working Group on the History of Geographic Thought* (W8), Autonomous University of Barcelona. August, 1986. 7 pp.
169	1986	K	"The Status and Stature of Raoul Blanchard: Glances from an Iberian Shore," *Working Group on the History of Geographic Thought*, International Geographical Union and International Union of the History and Philosophy of Science, Regional Conference on Mediterranean Countries. Autonomous University of Barcelona. August, 1986. 17 pp.
170	1987	B	"Subjective Experience and Objective Knowledge: Geography as Departure and Return Destination" in *Professor P. Pandey Commemorative Volume*, Post-Graduate Department of Geography, Ranchi University, Ranchi - 834001, India. (D.P. Satpathi, Editor.) 1987. 36 pp.
171	1987	I	"The Future of Geography." *Environments: A Journal of Interdisciplinary Studies*. Vol. 19. No. 1. May, 1987. (Review of a volume, edited by R.J. Johnston, Methuen, London and N.Y. 342 pp.)
172	1987	I	"Building for Music." *Environments: A Journal of Interdisciplinary Studies*. Vol. 19. No. 2. August, 1987. (Review of "Building for Music: The Architect, the Musician, and the Listener from the

Seventeenth Century to the Present Day." The M.I.T. Press, Cambridge, Mass. 371 pp.)

173 1987 G "Morrill's Inappropriate, Unwarranted Lardoons." *The Professional Geographer*. Vol. 39. No. 2. May, 1987. pp. 204-205.

174 1987 D "Geographers and the Future of Geography." *Transition*: *Quarterly Journal of the Socially and Ecologically Responsible Geographers*. Spring 1987. Vol. 16. No. 1. pp. 6-9. (Elaboration of Bibliography Item #171.)

175 1987 K "Epigramming Epic and Ephemeral Epicentral Epistemologies." *Acta Universitatis Lodziensis*. Folia Geographica 8 (1987) Łódź. Uniwersytet Łódzki. Łódź. 1987.

176 1987 B "Urban Geography, Regional Planning and Environmental Studies: A Synergistic Growth," *Récherches de Géographie Urbaine* (Hommage au Professeur J.A. Sporck.) Presses Universitaires de Liège. 1987. Vol. II. Part 5. (Numéro hors série du Bulletin de la Societé Géographique de Liège.)

177 1987 I "An Aesthetic Introduction to the Avoidance of Extinction," *Geojournal*: *International Journal for Physical, Biological, and Human Geosciences*. Akademische Verlagsgesellschaft. Wiesbaden. F.R. Germany. 0530-B. (Review of volume on "Vanishing Animals" by A. Warhol and K. Benirschke.)

University of Waterloo
Department of Geography Publication Series

Available from Publications, Department of Geography, University of Waterloo, Waterloo, Ontario, N2L 3Gl

28 DOUFOURNAUD, Christian and DUDYCHA, Douglas, 1987, *Waterloo Lectures in Geography, Vol. 3, Quantitative Analysis in Geography*, ISBN 0-921083-24-6, 140 pp.

27 NELSON, J. Gordon and KNIGHT, K. Drew, 1987, *Research, Resources and the Environment in Third World Development*, ISBN 0-921083-23-8, 218 pp.

26 WALKER, David F., 1987, *Manufacturing in Kitchener-Waterloo: A Long-Term Perspective*, ISBN 0-921083-22-X, 220 pp.

25 GUELKE, Leonard, 1986, *Waterloo Lectures in Geography, Vol. 2, Geography and Humanistic Knowledge*, ISBN 0-921083-21-1, 101 pp.

24 BASTEDO, Jamie, D. 1986, *An ABC Resource Survey Method for Environmentally Significant Areas with Special Reference to Biotic Surveys in Canada's North*, ISBN 0-921083-20-3, 135 pp.

23 BRYANT, Christopher, R., 1984, *Waterloo Lectures in Geography, Vol. 1, Regional Economic Development*, ISBN 0-921083-19-X, 115 pp.

22 KNAPPER, Christopher, GERTLER, Leonard, and WALL, Geoffrey, 1983, *Energy, Recreation and the Urban Field*, ISBN 0-921083-18-1, 89 pp.

21 DUDYCHA, Douglas J., SMITH, Stephen L.J., STEWART, Terry O., and McPHERSON, Barry D., 1983, *The Canadian Atlas of Recreation and Exercise*, ISBN 0-921083-17-3, 61 pp.

20 MITCHELL, Bruce, and GARDNER, James S., 1983, *River Basin Management: Canadian Experiences*, ISBN 0-921083-16-5, 443 pp.

19 GARDNER, James S., SMITH, Daniel J., and DESLOGES, Joseph R., 1983, *The Dynamic Geomorphology of the Mt. Rae Area: A High Mountain Region in Southwestern Alberta*, ISBN 0-921083-15-7, 237 pp.

18 BRYANT, Christopher R., 1982, *The Rural Real Estate Market: Geographical Patterns of Structure and Change in an Urban Fringe Environment*, ISBN 0-921083-14-9, 153 pp.

17 WALL, Geoffrey, and KNAPPER, Christopher, 1981, *Tutankhamun in Toronto*, ISBN 0-921083-13-0, 113 pp.

16 WALKER, David F., editor, 1980, *The Human Dimension in Industrial Development*, ISBN 0-921083-12-2, 124 pp.

15 PRESTON, Richard E., and RUSSWURM, Lorne H., editors, 1980, *Essays on Canadian Urban Process and Form II*, 505 pp. (Available only in microfiche).

14 WALL, Geoffrey, editor, 1979, *Recreational Land Use in Southern Ontario*, ISBN 0-921083-11-4, 374 pp.

13 MITCHELL, Bruce, GARDNER, James S., COOK, Robert, and VEALE, Barbara, 1978, *Physical Adjustments and Institutional Arrangements for the Urban Flood Hazard: Grand River Watershed*, ISBN 0-921083-10-6, 142 pp.

12 NELSON, J. Gordon., NEEDHAM, Roger D., and MANN, Donald, editors, 1978, *International Experience with National Parks and Related Reserves*, ISBN 0-921083-09-2, 624 pp.

11 WALL, Geoffrey, and WRIGHT, Cynthia, 1977, *Environmental Impact of Outdoor Recreation*, ISBN 0-921083-08-4, 69 pp.

10 RUSSWURM, Lorne H., PRESTON, Richard E., and MARTIN, Larry R.G., 1977, *Essays on Canadian Urban Process and Form*, 377 pp. (Available only in microfiche).

9 HYMA, B., and RAMESH, A., 1977, *Cholera and Malaria Incidence in Tamil, Nadu, India: Case Studies in Medical Geography*, 322 pp. (Available only in microfiche).

8 WALKER, David F., editor, 1977, *Industrial Services*, ISBN 0-921083-07-6, 107 pp.

7 BOYER, Jeanette C., 1977, *Human Response to Frost Hazards in the Orchard Industry, Okanagan Valley, British Columbia*, 207 pp. (Available only in microfiche).

6 BULLOCK, Ronald A., 1975, *Ndeiya, Kikuyu Frontier: The Kenya Land Problem in Microcosm*, ISBN 0-921083-06-8, 144 pp.

5 MITCHELL, Bruce, editor, 1975, *Institutional Arrangements for Water Management: Canadian Experiences*, 225 pp. (Available only in microfiche).

4 PATRICK, Richard A., 1975, *Political Geography and the Cyprus Conflict: 1963-1971*, 481 pp. (Available only in microfiche).

3 WALKER, David F., and BATER, James H., editors, 1974, *Industrial Development in Southern Ontario: Selected Essays*, 306 pp. (Available only in microfiche).

2 PRESTON, Richard E., editor, 1973, *Applied Geography and the Human Environment: Proceedings of the Fifth International Meeting, Commission on Applied Geography, International Geographical Union*, 397 pp. (Available only in microfiche).

1 McLELLAN, Alexander G., editor, 1971, *The Waterloo County Area: Selected Geographical Essays*, ISBN 0-921083-04-1, 316 pp.